Causation, Explanation, and the Metaphysics of Aspect

Causation, Explanation, and the Metaphysics of Aspect

Bradford Skow

OXFORD
UNIVERSITY PRESS

OXFORD
UNIVERSITY PRESS

Great Clarendon Street, Oxford, OX2 6DP,
United Kingdom

Oxford University Press is a department of the University of Oxford.
It furthers the University's objective of excellence in research, scholarship,
and education by publishing worldwide. Oxford is a registered trade mark of
Oxford University Press in the UK and in certain other countries

First Edition published in 2018

Impression: 1

Published in the United States of America by Oxford University Press
198 Madison Avenue, New York, NY 10016, United States of America

British Library Cataloguing in Publication Data
Data available

Library of Congress Control Number: 2018951949

ISBN 978–0–19–882696–5

Printed and bound in Great Britain by
Clays Ltd, Elcograf S.p.A.

For Deanna, and Iris, and Elliot, and Nathaniel

Contents

Acknowledgments

Thanks to audiences at MIT, Yale, and Berkeley for their attention and questions. Thanks to the two anonymous referees Oxford University Press recruited for writing such great referee reports. Thanks to Nathaniel Baron-Schmitt, Michael Della Rocca, Robin Dembroff, Nina Emery, Brian Epstein, Sally Haslanger, Sabine Iatridou, Harvey Lederman, Daniel Muñoz, Kieran Setiya, Zoltán Szabó, and Steve Yablo for reading and/or discussing this material with me. My sister Katy used some of the leisure time allowed to her by her computer science PhD program to paint the cover art; thanks Katy! Thanks to Peter Momtchiloff for continuing to support my work. As always, thanks to Fred Feldman for reminding me what good philosophy should do.

1

Advertisement

I

"Causation, explanation, okay. But, the metaphysics of aspect? What's that?"

Great question! In honor of the Monty Python boys, I'll start with something completely different. J. L. Austin, in *Sense and Sensibilia*, observed that something can exist without being real, since, for example, toy ducks certainly exist, even though they are not real ducks. This comment was part of Austin's ordinary-languaging of the word "real," but it sparked in his mind some thoughts about "exist," thoughts that, since they weren't directly relevant to his topic, he confined to a footnote:

> "Exist," of course, is itself extremely tricky. The word is a verb, but it does not describe something that things do all the time, like breathing, only quieter—ticking over, as it were, in a metaphysical sort of way. (Austin 1962, 68)

This "ticking over" bit is one of the great rhetorical moments in twentieth-century philosophy, but its greatness makes it easy to overlook what the footnote is missing. People usually quote Austin approvingly.[1] They agree with his claim. But Austin just asserts it. He doesn't

[1] An example: (van Inwagen 2009, 477). The temptation to make the mistake Austin opposed starts early. My older son asked me the other day "Do you know what I'm doing?" He was lying on the floor waving his hand in the air. "No, what?" "Existing."

argue for it. Once you realize this, so many questions are immediately urgent: why, exactly, doesn't "exist" describe something things do? What is it to do something anyway? What does *doing something* contrast with, and what can be said, in general, about which things fall on which side of the line?

I'll answer some of these questions in a minute, but first I want to change the subject one more time. Suppose that I strike a match and it lights, and that the match wouldn't have lit if I hadn't struck it. Now the match also wouldn't have lit if there hadn't been oxygen in the room. But, many want to say—I want to say—that only the striking is a cause of the lighting; the presence of oxygen is instead a "background condition" to the lighting. But then what is the difference between causes and background conditions? What did the striking do, that the presence of oxygen failed to do, that earned the striking the status of a cause? One tempting answer is that causes have to be events. But this answer only has something going for it if the striking of the match is an event and the presence of oxygen is not. Could that be right? This claim is easier to defend if it comes from some systematic theory of events, rather than being just a one-off judgment. So what general criteria could we state for being an event that the striking would satisfy and the presence of oxygen would not?

These two batteries of questions may seem unconnected, but I think that there is a single distinction that can help with both. Perhaps surprisingly, it is a linguistic distinction: the distinction between "stative" and "non-stative" verbs. The right way to draw the "did something"/"didn't do anything" distinction, and the right way to draw the event/non-event distinction, I am going to claim, uses the stative/non-stative distinction. Among other benefits to drawing these distinctions the way I will draw them is that the claims that drove the questions I started with come out true: existing will turn out not to be a way of doing something, and the presence of oxygen not to be an event.

The stative/non-stative distinction belongs to what linguists call the study of lexical aspect. The other two distinctions belong to metaphysics. The claim I will make, that the metaphysical distinctions "line up" with the aspectual distinction, is, then, a claim about the "metaphysics" of aspect.

II

This book consists of four essays built around three ideas. Two of the ideas are the ideas about the metaphysics of aspect I just mentioned (ideas I haven't actually stated in any detail yet—I'm about to get to that); the third is an idea about explanation. Each essay builds on the ideas in a different way, so I've written them so that they may be read independently. (This does make for some repetition.) My goal in this chapter is to introduce the ideas and the essays.

Before getting any metaphysics out of the stative/non-stative distinction we need more exposure to the distinction itself. Which verbs are stative, and which are non-stative? One test uses the distinction between a progressive clause, like "Jones was singing," and a non-progressive clause, like "Jones sang" (a distinction belonging to "grammatical" aspect). For the most part, non-stative verbs may appear in the progresive while stative verbs may not. For example "paddle" is non-stative, and may appear in the progressive: "Jones was paddling the boat" is grammatical. "Be" (the "be of predication," as in "Jones is tall") on the other hand, is stative, and may not appear in the progressive: "Jones is being tall" is ungrammatical. This test for stativity is not perfect—stative verbs can sometimes appear in the progressive. Fortunately there are other tests, one of which I'll discuss in a minute.[2]

[2] These are tests for when an English verb is stative. The stative/non-stative contrast exists in other languages too. But I don't know enough to say anything about them. (Sabine Iatridou tells me that the translation of the "be" of predication into some languages can be non-stative, when it bears perfective marking, something it cannot do in English.)

Comrie, in his book *Aspect*, discusses both the progressive/non-progressive and stative/non-stative distinction; he also discusses progressive uses of stative verbs (Comrie 1976, 37–9).

How does the stative/non-stative distinction help with the question of when an event exists? It is easy to give examples of conditions under which there exists an event, harder to produce a principled theory that fits them. If a bomb explodes, then in virtue of this fact an event occurs, namely the explosion of the bomb. By contrast, even though $2+2=4$, it is false that in virtue of this fact an event ("2 plus 2's being 4") occurs. What is the difference between these two cases? More generally, can we say anything useful and informative about which clauses can go in for "S" to make the following true?

If S, then an event occurs in virtue of the fact that S.

I hold that you get a truth from the above schema when you put in for "S" a simple clause whose main verb is non-stative.[3]

[3] The clause must be in the present tense and non-progressive. I also endorse the past tense and the progressive analogues of this schema.

This answer is not mine originally. It is a key plank in the "Neo-Davidsonian" approach to the semantics of natural language. Parsons (1990) defends the approach and discusses its history. It gets its name from some papers by Donald Davidson, collected in (Davidson 2001).

In chapter 5 I end up restricting this claim to exclude non-stative verb phrases headed by raising verbs; since "fail" is a raising verb, with this restriction the theory does not entail that there are "negative events" like failures. See that chapter for my reasons for doing this, and a definition of "raising verb."

It may be that I should make even more restrictions. When Socrates died Xanthippe became a widow. If "become a widow" is non-stative, then my theory says that there was a corresponding event. But many philosophers have thought that while an event (a death) occurred in virtue of the fact that Socrates died, no event occurred in virtue of the fact that Xanthippe became a widow. I'm inclined to agree. My main reason, which I'll discuss later, is that I want a theory of events that complements the thesis that events are things that can cause or be caused. "Xanthippe's becoming a widow" doesn't seem like it can cause anything, or be caused by anything. Now I'm a bit unsure about whether or not "become a widow" is non-stative, but let's suppose it is. Then in outline I want to restrict the true instances of "If S, then an event occurs in virtue of the fact that S" to exclude instances in which what goes in for "S" has as its main verb phrase a verb phrase like "become a widow." The problem, of course, is saying which verb phrases those are. I'm afraid I don't know how to do this. (One suggestion is that they are verb phrases that

This theory gets the cases right. The verb "explode" is non-stative: "The bomb was exploding" is grammatical. So if a bomb explodes, an event (an explosion) occurs in virtue of this fact, as we want. But the "is" of identity is stative: "2 plus 2 is being 4" is ungrammatical. So no event occurs in virtue of the fact that $2 + 2 = 4$. It also gets the examples that motivate the "cause/background condition" distinction right. The theory entails that an event occurred in virtue of the fact that I struck the match, but no event occurred in virtue of the fact that oxygen was present in the room.

If this theory is right, then it is natural to ask whether there are things whose existence "goes with" the truth of sentences with stative verbs in the way that (according to this theory) the existence of events goes with the truth of sentences with non-stative verbs. I say that there are, namely states. If oxygen is present in a room, then in virtue of this fact a state obtains. States, however, are different from events, in a variety of ways; for one thing, only events can cause or be caused. States cannot.[4] (These claims come in for more attention in chapter 5.)

For many non-stative verbs—this is a bit of a digression—there is a noun spelled the same as the present-participle form of that verb, a noun that applies to the corresponding event. "Stab" is non-stative, and the noun "stabbing" (as in "three stabbings") applies to events that happen when something of the form "X stabs Y" is true. So when there is such a noun we can go beyond the bare claim that when "X Ved" is true and V is non-stative then there was a corresponding event;

denote mere changes in something's relations to other things. But that can't be right, since—I hold—motion is a mere change in something's relations to other things, yet when something moves there is a corresponding event.) For the most part examples like this won't come up, so for the most part I will ignore this problem.

[4] The word "state" is reminiscent of the phrase "state of affairs," a phrase many philosophers have used to name one kind of thing or another (I'm not sure whether they've always used it to name the same kind of thing). But I'm not using "state" to abbreviate "state of affairs," so please don't assume that the properties other philosophers have said states of affairs have are properties I say states have. I realize that this means I say very little about the properties states have. I think I say enough about them for the work I want states to do.

we can say that the event was a Ving. But there is not always such a noun.[5] Still, I will sometimes pretend that there is when I want to state generalizations about facts statable using non-stative verbs and the events that occur in virtue of them. (I'm about to do this in the next sentence.)

The claim that a Ving is happening whenever something is Ving follows from, but is weaker than, the claim that "X is Ving" means the same as "An event that is a Ving by X is happening." Partisans of "Neo-Davidsonian semantics" endorse something like this stronger thesis: they hold that non-stative verb phrases, and only non-stative verb phrases, are actually (at the level of "logical form") predicates of events (see Parsons 1990). But you don't need to be a Neo-Davidsonian to think that non-stative verbs go with events. You can hold that "Jones is crossing the street" entails that a crossing is happening without meaning "A crossing of the street by Jones is happening." You could be even more cautious and hold that it entails this only in conjunction with the assumption that there are such things as events.

[5] "Announce," for example, is non-stative, so if Jones announced that Smith had won, a corresponding event occurred. But this event was not "an announcing that Smith had won"—this phrase is not grammatical. (Of course we do have a noun for this kind of event: it was an announcement.)

Huddleston calls nouns that share a shape with a present-participle form of a verb "gerundial nouns" (2002, 81–2). In some cases, the gerundial noun derived from a non-stative verb phrase is a mass noun, not a count noun. "Run" is an example (as is any so-called "activity" verb phrase): it's not right to say that when Jones ran, there was a running by Jones. (Mass nouns cannot take determiners like "a" or "three.") I don't know what to make of the fact that some non-statives correspond to count nouns and some to mass nouns. Presumably events are things that can be counted, so it seems that non-statives that correspond to mass nouns do not "go with" events or with states. Mourelatos (1978) claims that non-statives that correspond to mass nouns go with processes rather than events. I still maintain, however, that the stuff that activity verbs go with is in some way "event-like," and so should be grouped with events and not with states. To save space in the main text, but only for this reason, I'll often write as if the gerundial noun derived from a non-stative is always a count noun.

Okay, I've presented a partial theory of events. It is far from a complete theory. It doesn't say anything about the "identity conditions" of events. If Jones crossed the street slowly, he crossed the street, and since "cross" is non-stative the theory says that there was a slow crossing by Jones, and also a crossing by Jones. But the theory doesn't say anything about whether there was just one crossing, or more. The theory also doesn't say anything about the essences of events. It doesn't say whether Jones' crossing is a crossing in every possible world in which it happens. For all the theory says, the property of being a crossing is a property that event has contingently.[6]

While it is far from complete, the theory does make a substantive claim about the conditions under which an event is occurring. So why believe it? It gets the small number of cases I've looked at right. But so do other theories. So does, for example, the theory that an event happens whenever something changes. When a bomb explodes, the bomb changes, but when $2+2=4$ nothing changes. Yet these are distinct theories: a non-stative verb phrase can apply to you even if you don't change. "Stand still," for example, is non-stative, but someone who is standing still need not be changing.

Why prefer my theory to the theory that "events are changes"? Some may intuit that the claim

> when someone stands still an event occurs in virtue of this fact

is false. I myself don't share that intuition. More importantly, I am interested in arguments for and against my view that don't rest simply on intuitions about when an event does or does not happen. This is because I want a theory of events with the following two features. First, the line it draws between events and states should be a natural one, a

[6] Lots of philosophers have weighed in on the identity conditions and essences of events. For just a few references, Bennett seems to hold that when Jones crossed the street then there was just one crossing, which was slow (Bennett 1988), while Yablo holds that there were at least two crossings, both of them slow, one of them essentially slow, the other not (Yablo 2010).

"joint in nature." Second, the theory should "play well" with the thesis that events can be causes and effects while states cannot. And whether the theory gets intuitions about when an event happens by and large right does not seem to me to bear very directly on whether the theory has either of these features. I say something in defense of the theory's having these two features in chapters 2 and 5.[7]

I've said a lot about events; I need to get back to J. L. Austin, "exists," and doing something. He said that "exists" does not describe something things do, but left us asking what demarcates doing something from its contrary. Now it will be useful to have an abbreviation for "X did something," and I will use "X acted" for this purpose. I think this is a perfectly good thing, and a perfectly ordinary thing, to mean by "X acted." But to forestall confusion I should say that this is not the only meaning "acted" has in philosophy. Some philosophers use "X acted" to mean what I mean by "X acted intentionally" or "X acted for a reason"; if Jones sneezed involuntarily he did not act, in the more demanding sense, but did act, in my less demanding sense, since he did something, namely sneeze.[8]

Okay, now as with the event/non-event contrast it is easy to give examples of the act/non-act contrast. If Jones paddled his canoe, he

[7] Neo-Davidsonians will argue that the theory I've written down follows from their semantic theory, and so any argument for their semantic theory is an argument for the theory I've written down. I won't look a gift horse in the mouth; I'm happy to endorse arguments like that. But I'm not going to be giving any in this book.

[8] Davidson's use of "action" in his work (Davidson 2001), and the literature that engages with it, is related to the more demanding sense of "act": for him, an action is an event that happens when someone acts (that is, does something) intentionally.

It was Kieran Setiya who taught me that action theorists often use "act" to mean "act for reasons," in (Setiya 2009); in that paper he also emphasizes the importance of recognizing the broader sense of "act," and the connection between acts and non-stative verbs. (These themes play a role in several of Setiya's papers; the Introduction to (Setiya 2017) has a summary presentation.) It was reading Setiya's work that got me thinking about this connection. That thinking led to this book. Kieran: I tip my hat to you.

thereby did something. But if Jones was six feet tall, it is false that he thereby did something. What is the difference?

This question has several readings. On some ambitious readings, it asks for the features that are definitional of doing something, or for the features that are uniquely essential to doing something (in the sense that they are not also all essential to something else). I wish I had answers to these questions, but I don't. I'm going to answer the question on a less ambitious reading, a reading on which it asks: is there some informative generalization that separates acting from not acting? Is there some informative condition that can go on the right-hand side of the following to get a truth?

> If X Ved, then in virtue of this fact X did something if and only if...

My answer is that you get a truth if you put "'V' is non-stative and in the active voice" on the right:

> If X Ved, then in virtue of this fact X did something if and only if "Ved" is a non-stative verb phrase in the active voice.

An argument for this thesis goes like this. The first, and central, premise says that "One thing X did was V" is grammatical iff V is non-stative (and appears in the sentence in its plain form). You can convince yourself of this premise by checking instances. For example, "One thing Jones did was break the window" is grammatical while "One thing Jones did was be tall" is not, and "break" is non-stative while "be tall" is stative. The argument for the right-to-left direction then goes like this: suppose (i) that V is non-stative, (ii) that it occurs in the active voice in "X Ved," and (iii) that it is true that X Ved. Now if "One thing X did was V" is grammatical, then "X Ved" entails it if "Ved" is active.[9] By (i) this is grammatical. By (ii) and

[9] For example, "stab" is non-stative, and the active "Jones stabbed Smith" entails "One thing Jones did was stab Smith," but the passive "Smith was stabbed by Jones" does not entail "One thing Smith did was be stabbed by Jones" (in fact this last sentence is not grammatical).

(iii) it is true. So one thing X did was V. But obviously "one thing X did was V" entails "X did something." Now since all this reasoning was conducted under the supposition that X Ved, if we discharge the supposition we have that if X Ved, then X did something. I think it clear that it also establishes that if X Ved, then *in virtue of this fact* X did something (though I can't prove this using a theory of the "logic" of this use of "in virtue of"). The argument for the other direction is similar.

This argument is only as strong as its premise, and there is a problem with the premise: I haven't given you a perfectly reliable test for non-stativity. Well I can give you a perfectly reliable test: V is non-stative iff "One thing X did was V" is grammatical (when V is put in its plain form).[10] But using this test to support the premise would beg the question.

This problem with the argument is not important. I am in fact happy to dispense with argument and just define "non-stative verb phrase" as "verb phrase that, when put into its plain form and inserted into 'One thing X did was…,' yields a grammatical sentence."[11] My "insight" about aspect now becomes "true by definition." This is okay: in this book what will matter is that it is true that acting goes with non-stative verb phrases, not that this is a "substantive" truth.

One might object that whether someone has acted doesn't in any way depend on language, but that my claim does make it depend on language. In fact that's not so. My claim might be in danger of making whether someone has acted depend on language if it entails that you can't do something without an active non-stative verb phrase being

[10] Szabó regards this as the best test for non-stativity (Szabó 1994), though this fact is compatible with his thinking it not perfectly reliable. In the cited paper he discusses its relation to the "can occur in the progressive" test.

[11] That is, I am happy to do this for the purposes of this book. I think that a non-question-begging argument that this test is perfectly reliable could be given, but that doing so would involve going into more detail about what linguists say about non-stativity than would be fruitful here.

true of you. But it doesn't entail this.[12] If you do something, but the language lacks a verb phrase that describes your situation, then you've done something without there being any non-stative verb phrase that is true of you.

In fact, as my remarks in the previous paragraph suggest, I am tempted by the view that the dependence goes the other way: whether something is a non-stative verb phrase depends on whether it denotes an act. This view answers a question you may have had since I introduced the notion of a non-stative verb phrase. My criteria for being non-stative are all syntactic. Is there a criterion that sorts the stative verbs from the non-stative ones in terms of their meanings? Yes (on this view): the non-stative verbs denote acts.

To wrap things up, we now have an argument where Austin had none: "exist" is a stative verb, so no, it does not describe something things do. After all, if it did, we could say that one thing I did yesterday was exist, when in fact this is not true (or false), because it has a stative verb where only a non-stative verb may go.

III

Back to events: the claim that the occurrence of an event goes with the truth of a clause with a non-stative verb may get various examples right, but it can feel like it comes out of left field. One may hesitate to accept that the claim is true without some answer to the question of why it is true. The connection between non-stativity and acting is a step toward an answer. An obvious consequence of the fact that the truth of a clause with a non-stative verb goes both with the occurrence

[12] I'm not saying that the theory makes acting depend on language if it *does* entail this. "X did something only if a non-stative verb phrase is true of X" does not by itself entail that whether X did something *depends on* whether a non-stative verb phrase is true of X. But it does leave the path open for this dependence claim to be true.

of an event and with acting is that the occurence of an event goes with acting. The consequence is that if something did something, then in virtue of that fact an event occurred; and if an event occurred, that's so in virtue of the fact that something did something. If this consequence is true then it explains why events go with non-stative verbs. And the consequence is plausible: surely it is plausible that when something does something that's enough for an event to occur.

The other direction is maybe less plausible: if an event happens, must that be in virtue of something's doing something? But it is harder to find counterexamples than you might think. Some have said that raining is an event, but that when it rains, nothing is doing anything. If anything is doing anything when it rains, it must be the thing denoted by "it" in "it is raining"; but "it" is here a "dummy" pronoun, not denoting anything. Right? Well, I'm not so sure (and can't find a consensus among linguists[13]). Anyway, the syntax and semantics of "it is raining" aside, I think that there is something doing the raining, namely the rain.[14]

IV

I mentioned earlier that the chapters in this book do not constitute a single argument, each building on the last. Instead they share a common starting point, one that comprises the two insights about the metaphysics of aspect—the connection to events and the connection to acts—and a third insight, about levels of explanation. In any given chapter one or another insight may be closer to the surface and the others more submerged; then in the next chapter the roles might switch. But they're all there in the second chapter, "A Theory of

[13] The wikipedia page for dummy pronouns <https://en.wikipedia.org/wiki/Dummy_pronoun> lists some "dissenting views," which include the views of Noam Chomsky.

[14] Unless it's raining cats and dogs. Then it is the cats and dogs that are raining down.

Background Conditions." That chapter is also the place where the two ideas about aspect get their most sustained discussion.

I want to say a bit about where each chapter is heading, but first I need to say what the insight about levels of explanation is. The first thing I want to say is that the insight is really better termed an insight about levels of reasons. What's the difference? An explanation, or, at least, the kind of explanation philosophers have been interested in theorizing about, is just an answer to a why-question. And reasons are the "basic parts" of answers to why-questions, and so are the basic parts of explanations. When someone tells you that "part of why" rents are so high is that it has become harder to get a mortgage, they are telling you one of the reasons why rents are high; when they tell you all of the reasons, they have given you the complete answer to the question of why rents are high.[15]

So what do I mean by "levels of reasons"? I have in mind the distinction between the reasons why Q, on the one hand, and the reasons why those facts are reasons why Q, on the other. I call reasons of the first kind, first-level reasons, and reasons of the second kind, second-level reasons. If I strike a match and thereby cause it to light, one reason why the match lit is that I struck it; that I struck the match is a first-level reason with respect to the lighting. Among the second-level reasons is the fact that oxygen was present; this fact is part of the answer to "Why is it that the fact that I struck it is a reason why the match lit?"

After all this stage-setting, what is the insight? My central thesis about levels of reasons is that *second-level reasons need not also be first-level reasons.* If R is a reason why Q, not every reason why ⟨R is

[15] I defended these claims, and the claim about levels of reasons that I am about to make, in *Reasons Why* (Skow 2016). In that book I worked very hard to avoid the word "explanation." I believed, and still believe, that building your theory using that word is apt to lead you down a blind alley, and that you can avoid a lot of trouble by just always speaking of why-questions and their answers instead. But like a chocolate-lover who abstains for just a week and thinks he's proved he can live without it, I've let the word slip back into my writing.

a reason why Q⟩ is also itself a reason why Q. I worked hard in my last book to strip away facts that various philosophers have held were first-level reasons and put them where I think they belong: the place of second-level reasons. For example, Carl Hempel famously held that every explanation must cite a law of nature (Hempel 1965). Translated into the language of reasons his thesis is that whenever there are any reasons why Q, at least one of those reasons is a law of nature. I deny this, at least in the case of reasons why an event happened. For a given event E, no law of nature is a reason why E happened. But laws are still indirectly relevant to why E happened, because some law is a reason why something else (a cause of E for example) is a reason why E happened. If I drop a rock from 1 meter above the ground, and it hits the ground at 4.4 meters per second, one reason why it hit at that speed is that I dropped it from that height. The law that impact speed s is related to drop height d by the equation $s = \sqrt{2dg}$, however, is not a reason why the rock hit the ground at 4.4 m/s.[16] It is, instead, a reason why ⟨that I dropped the rock from 1m up is a reason why it hit the ground at 4.4 m/s⟩.

According to an old proverb, if all you have is a hammer, everything looks like a nail. The distinction between levels of reasons is my hammer, but when I published *Reasons Why* I didn't appreciate how many nails were lying about. Throughout these chapters I apply the distinction to new areas.

In "A Theory of Background Conditions" (chapter 2) I apply the distinction between levels of reasons to background conditions, and do to them what in *Reasons Why* I did to laws. I argue that a background condition to C's causing E is not both a reason why E happened and a reason why C's happening is a reason why E happened; instead it is just a second-level reason. The presence of oxygen, for example, is not both a reason why the match lit and a reason why my striking the match is a reason why the match lit; it is just a reason why my striking the match is a reason why the match lit.

[16] The constant g is the gravitational acceleration near the surface of the earth; the law is valid only when the drop height is small and drag is negligible.

As a consequence, dispositions also "move up a level." When a fragile glass is struck and breaks, that it was fragile is not a reason why it broke; instead, that it was fragile is a reason why ⟨that it was struck is a reason why it broke⟩. This follows because, as I discuss in the chapter, when something manifests a disposition, that it has the disposition is a background condition to the manifesting of the disposition.

These claims contradict claims I made in *Reasons Why*. There I said that background conditions were reasons of both levels. That was because I couldn't see a difference between causes and background conditions that could justify asserting that causes were first-level reasons (with regard to their effects) while denying that status to background conditions. A common view is that background conditions are "metaphysically" no different from causes, the difference is just pragmatic: background conditions are causes that we are ignoring in our conversation. If that's right, then since on my view all causes (speaking unrestrictedly) are reasons why their effects happen, background conditions too must be reasons why the relevant effects happen.

The problem with this argument is with the common view. It is false. Background conditions *are* metaphysically different from causes: they are states, while causes must be events. So background conditions are not causes, not even unrestrictedly speaking. On my view this prevents them from being reasons why the relevant events happened. Something can be a reason why E happened only by being a cause of E or by being a ground of E's happening (and a background condition to C's causing E is certainly not a ground of E's happening).

Now the idea that background conditions are second-level reasons should not be seen as a convenient place to retreat to, once you've been chased off the appealing idea that they are first-level reasons. It's a place you'd want to plant your flag to begin with. Why? I am not the first to say that background conditions are states rather than events. But saying this is no more a theory of background conditions than "causes are events not states" is a theory of causation. What does a state have to be like to be a background condition to the occurrence of a given event E? This is the question that needs to be answered, and the correct answer must not make background conditions so much like causes that the distinction between them is metaphysically uninteresting. My idea is what we need: a state is a background condition to E when it is a reason why something else is a cause of E; in this, background conditions differ drastically from causes of E, which are in a direct way reasons why E happened.

A second ago I slid from the claim that a background condition to C's causing E is a reason why C's happening is a reason why E happened, to the claim that it is a reason why C is a cause of E. I happen to think that a background condition is both, and that this is not a coincidence; in general, anything that is a reason why C is a cause is also a reason why C is a reason, and is a reason why C is a reason because it is a reason why C is a cause (see Skow 2016, chapter 5). In light of this I'll sometimes blur the line between a reason why something is a reason (a higher-level reason) and a reason why something is a cause.

The engine that pushes background conditions up to become second-level reasons is the claim that background conditions must be states. Here is where the idea that states go with stative verbs is important. If you're going to say that background conditions must be states, you're going to need a theory of events and of states that at least says that the presence of oxygen in the room is a state, not an event. Not every theory says this: many philosophers accept theories of events according to which the presence of oxygen is an event. But that's wrong. My theory says the right thing. (In the chapter I say more about what I think is wrong with one competitor to my theory of events, namely the "property instantiation" theory of events.)

The other metaphysical correlate of the stative/non-stative distinction, the distinction between acts and non-acts, takes center stage in "Dispositions: Intrinsicness and Agency" (chapter 3). Launching that chapter is the observation that "X is disposed to Z" is only grammatical if "Z" is a non-stative verb phrase in the active voice. A disposition, therefore, must be a disposition to act. I use this fact to explore the dispute over whether there can be extrinsic dispositions. Many alleged examples of extrinsic dispositions fail, not because the disposition in the example is actually intrinsic, but because there is no disposition in the example. Jennifer McKitrick suggests, for instance, that the disposition to be seen if looked at is extrinsic (McKitrick 2003); but there is no such disposition, since "be seen" is not a non-stative verb phrase. "Be seen" is not something one can do.

The claim that dispositions must be dispositions to act does not save the thesis that dispositions must be intrinsic from all potential counterexamples. Extrinsic dispositions do exist. But it does suggest a weaker thesis that might be true. If a disposition must be a disposition to act, then a familiar distinction between kinds of acts translates into a distinction between kinds of dispositions. Performances of acts divide into performances of basic acts and of non-basic acts. One does something as a non-basic act when one does it by doing something else; otherwise one does it as a basic act. So then if every disposition is a disposition to act, every disposition is either a disposition to do something as a basic act or a disposition to do something as a non-basic act. But if I'm disposed to do something as a non-basic act, I can only manifest the disposition by doing something else. Presumably in doing this other thing I also manifest some disposition. Let us say then that a non-basic disposition is a disposition you manifest by manifesting some other disposition, and that a basic disposition is one that is not non-basic.[17] Might the intrinsicness thesis be true of the basic dispositions? After all, it's false that every property is intrinsic, but much more plausible that every basic (that is, fundamental, or perfectly natural) property is intrinsic.[18] I end the chapter by discussing this weaker version of the intrinsicness thesis (worded slightly differently) and claiming that it is defensible.

[17] Of course there are other distinctions between dispositions that might merit the label "basic/non-basic." Some properties are more fundamental than others, closer to the "metaphysical ground floor"; one might want to use "basic disposition" for the most fundamental dispositions. As long as we keep in mind that we're using the same label for two distinctions we shouldn't get confused. (Also, it is not crazy to think that the distinctions might line up: shouldn't a disposition to do something as a basic act be more fundamental than a disposition to do something as a non-basic act?)

[18] It is a key plank of David Lewis's metaphysic that perfectly natural properties are intrinsic; see for example (Lewis 1983).

In “Structural Explanation: Garfinkelian Themes” (chapter 4) a moral we may draw from the distinction between levels of reasons is important: sometimes it is important to focus, not on an event’s causes, but on the reasons why those causes are causes. In the chapter I take up the question of what a structural explanation is, or might be. Structural explanations are important in the social sciences; they are often said to be better in some respect than “individualistic” explanations. But, I think, there is a lot of unclarity about what structural explanations are. I aim to isolate one kind of structural explanation that deserves more attention, and the notion of a reason why something is a cause plays a central role in my definition.

Now often when structural explanations are said to be superior to individualistic explanations, the two kinds of explanation are seen as competing forms of answer to one and the same why-question. To a single why-question, say the question of why Lisa quit her job while her husband Larry kept his, there is thought to be both an individualistic answer and a structural answer. One interesting feature of the kind of structural explanation I am interested in is that it does not compete with individualistic answers to why-questions. Instead, individualistic answers and structural answer are answers to different questions. In short, my view is this: you have a structural explanation when a structural fact (for example, a fact about the structure of society) is a reason why one thing (or kind of thing) causes another. With respect to Lisa and Larry, an individualistic answer to the question of why she quit and he didn’t is correct; she quit and he didn’t because they both believed that this was the choice that had the best consequences for the well-being of their family. The structural explanation answers the different question of *why her quitting and his not had certain effects* (in this case, on the well-being of their family). I pursue this theory of structural explanation through a reading of Alan Garfinkel’s book *Forms of Explanation* (1981).

My argument in chapter 2 that a background condition to E is not a reason why E happened depends on that claim that causes are events and background conditions are not. That causes are events is widely accepted; but why accept it? Why can’t states be causes or effects? I take up this question in “Agent Causation Done Right” (chapter 5).

I argue there that the most fundamental kind of causation is not causation by events, but causation by *things* (like people and rocks). More specifically, the most fundamental causal locution is "X caused Y to Z by Ving," where terms for things (that are not events) go in for "X" and "Y." Moreover—here comes the metaphysics of aspect again—instances of this schema are only true when the verbs that go in for "Z" and "Ving" are non-stative verbs. This entails that event causation makes sense, that is, that we can give truth-conditions for sentences of the form "C caused E," where terms for events go in for "C" and "E," using thing-causation. Since events go with non-stative verbs, the event C corresponds to a truth of the form "X Ved," where what goes in for "V" is non-stative; similarly E corresponds to "Y Zed." So "C caused E" is true iff X caused Y to Z by Ving. Since states correspond to stative verbs, no such treatment of state causation is possible. States cannot be causes.

I called the kind of causation involved when X causes Y to Z thing-causation to avoid saying too many controversial things at once, but really I prefer to call it "agent" causation. By this I do not mean what "agent causation" has come to mean in the debates over free will and determinism and in the debates over the nature of intentional action. As I use "agent causation" there are things that are "agent-causes," like rocks, that lack free will entirely and are not capable of acting intentionally. I call it agent causation because, if X causes Y to Z by Ving, then (obviously) X Ved, and since what goes in for "V" here is a non-stative verb, by Ving X acted and was thereby an agent. I argue in chapter 5 that there are a variety of puzzles and questions about causation that are easier to solve and answer if agent causation is basic than if event causation is basic.

V

The metaphysical questions about aspect I address in this book, and the problems to which I apply my answers, are the tip of an iceberg. I hope what I say encourages others to dive in and explore. There is so much more.

Here are just a few natural questions. First some stage-setting: some metaphysicians spend a lot of time wondering what the "fundamental language" for describing the world looks like.[19] They regard this not as a silly question about language, but as part of the profound question of what the world is like at the most fundamental level. On their view, you answer the profound question by figuring out which sentences in the fundamental language are true, and to do that, you need to be able to identify which sentences are sentences of that language in the first place.

The menu of questions people have asked about the fundamental language includes entries like, does it contain symbols for both conjunction and disjunction? Does it contain modal operators? Do its monadic predicates apply to things larger than a point of space(time)? But metaphysicians have by and large ignored questions like: are all the verbs (or predicates) in the fundamental language stative? Non-stative? Does the language have verbs of both kinds? But these questions are important. If the fundamental language has only stative verbs, then in a sense "the non-stative reduces to the stative." Every sentence with a non-stative verb can be given truth-conditions in which every verb is stative. Could that be right?

I would guess that these questions have been missed because Quine succeeded to some extent in getting metaphysicians to conduct their debates in formal languages that, like the language of first-order logic, lack a syntactic distinction between stative and non-stative predicates. But it would beg a lot of questions to take this as evidence that metaphysics should ignore the distinction. Maybe the absence of the distinction is a way in which those languages are impoverished compared to natural languages, not a way in which they are superior.[20]

[19] The most complete recent treatment of the question is (Sider 2011). Not all schools of meta-metaphysics think this is a good question.

[20] It is worth noting here that Galton, in *The Logic of Aspect* (Galton 1984), argues that standard presentations of tense logic are inadequate exactly because they lack aspectual distinctions. Quine advocated the use of formal languages in, for example, (Quine 1960).

VI

In "A Plea for Excuses" Austin said something about how he thought one should do philosophy: "we may use the dictionary...read the book through, listing all the words that seem relevant" to your question. Anticipating shocked responses, Austin assured us that "this does not take as long as many suppose" (1979, 186). Austin's method is, of course, crazy. Instead, we may use a grammar book. Read the book through. It may take a while but it's worth it. Anyway, if Austin had spent more time with grammar books than with dictionaries he might have had an argument to put in his footnote.[21]

References

Austin, J. L. 1962. *Sense and Sensibilia.* Oxford University Press.

Austin, J. L. 1979. "A Plea for Excuses." In J. O. Urmson and G. J. Warnock (eds.), *Philosophical Papers.* 3rd edition. Oxford University Press, 175–204.

Bennett, Jonathan 1988. *Events and Their Names.* Hackett.

Comrie, Bernard 1976. *Aspect.* Cambridge University Press.

Davidson, Donald 2001. *Chapters on Actions and Events.* 2nd edition. Oxford University Press.

Galton, Antony 1984. *The Logic of Aspect.* Oxford University Press.

Garfinkel, Alan 1981. *Forms of Explanation.* Yale University Press.

Hempel, Carl 1965. *Aspects of Scientific Explanation and Other Chapters in the Philosophy of Science.* The Free Press.

Huddleston, Rodney 2002. "The Verb." In Rodney Huddleston and Geoffrey K. Pullum (eds.), *The Cambridge Grammar of the English Language.* Cambridge University Press, 71–212.

[21] Okay, I don't actually know anything about how much time Austin spent with the various research materials available to him. Nor do I know if the grammar books he would have read even mention the stative/non-stative distinction. Austin died in 1960, when contemporary thinking about lexical aspect was only just getting started: the two works usually cited as originating that thinking are (Vendler 1957) and (Kenny 1963). For all I know, if the study of lexical aspect had been more advanced when Austin was working he would have had ideas about its philosophical importance that make mine look like left-handed crayon drawings.

Kenny, Antony 1963. *Action, Emotion and Will.* Routledge & Kegan Paul.
Lewis, David 1983. "New Work for a Theory of Universals." *Australasian Journal of Philosophy* 61: 343–77.
McKitrick, Jennifer 2003. "A Case for Extrinsic Dispositions." *Australasian Journal of Philosophy* 81: 155–74.
Mourelatos, Alexander 1978. "Events, Processes, and States." *Linguistics and Philosophy* 2: 415–34.
Parsons, Terence 1990. *Events in the Semantics of English.* MIT Press.
Quine, W. V. O. 1960. *Word and Object.* MIT Press.
Setiya, Kieran 2009. "Reasons and Causes." *European Journal of Philosophy* 19: 129–157.
Setiya, Kieran 2017. *Practical Knowledge: Selected Essays.* Oxford University Press.
Sider, Theodore 2011. *Writing the Book of the World.* Oxford University Press.
Skow, Bradford 2016. *Reasons Why.* Oxford University Press.
Szabó, Zoltán 1994. "On the Progressive and the Perfective." *Nous* 38: 29–59.
van Inwagen, Peter 2009. "Being, Existence, and Ontological Commitment." In David Chalmers, David Manley, and Ryan Wasserman (eds.), *Metametaphysics.* Oxford University Press, 472–506.
Vendler, Zeno 1957. "Verbs and Times." *The Philosophical Review* 56: 143–60.
Yablo, Stephen 2010. "Cause and Essence." In *Things.* Oxford University Press, 59–97.

2

A Theory of Background Conditions

I

The paradigm case of the contrast between a cause and a background condition is the contrast between the striking of a match and the presence of oxygen. I strike a match in a room filled with air and, by striking the match, light it. My striking of the match caused the match to light; the presence of oxygen in the room did not. Instead, it was a background condition to the lighting.

One common view about the cause/background condition distinction is that the distinction is drawn in language rather than the world; from the point of view of metaphysics, causes and background conditions are the same kind of thing. This view appears in, for example, J. L. Mackie's discussion of counterfactual theories of causation, in *The Cement of the Universe*. The simplest counterfactual theory says that X is a cause of Y iff, had X not happened, Y would not have. Mackie discusses three "difficulties" for this theory, the third of which concerns

> the distinction which we are inclined to draw between conditions and causes. There may be a set of factors which were, in the circumstances, jointly sufficient and severally necessary for a certain result, and which all occurred, as, consequently, did the result. Then we can say of each of these factors that if in the circumstances it had not occurred the result would not; but we may not be so willing to say of each

> of them that it caused the result. Perhaps we *ought* to say this, since they are all logically related to the result in the same way: the situation as described is symmetrical with respect to all the factors. (1980, 34)

Had there been no oxygen in the room, the match would not have lit, so the oxygen satisfies the counterfactual theory's criterion, yet we are "not so willing" to say that the presence of oxygen was a cause. Is this a counterexample to the counterfactual theory? Maybe not, Mackie goes on to suggest; maybe it *is* a cause (we "ought" to say it is), and our unwillingness to say it is a cause has some pragmatic explanation.

The explanation is pragmatic in the sense that it appeals to the role of conversational context. One of the context's roles is to help fix the semantic values of context-sensitive expressions; and the common view (as I will interpret it) says that both "cause" and "background condition" are context sensitive. To start with "cause," the common view says that this word is usually used in a restricted sense. There are various theories about how this restriction might work; one simple theory says that "C was a cause of E" is usually elliptical for "C was a cause of E *that X*" (for example, C was a cause of E *that was initiated by a human agent*, or C was a cause of E *that involved something changing*), and that the context in which the clause is used supplies the value for "X." The common view goes on to say that "C was a background condition" just means "C was a cause (unrestrictedly speaking), but not a cause that X." Background conditions are just causes that have been put in the "contextual" background. The common acceptance of "the presence of oxygen is not a cause, just a background condition" is then to be explained by saying that (i) people are commonly in contexts in which "cause" is being used to abbreviate something like "cause that involves a change of some kind," and (ii) the striking of the match involves change, while the presence of oxygen does not.[1]

[1] John Stuart Mill accepted something like the common view (Mill 1846, 198). More contemporary philosophers who accept something like it include

This common view is seductive because it is partly right. Its first claim is true: "cause" is often used in a restricted sense. News stories that say things like "There were three causes of the explosion" are using "cause" in a restricted sense; if there are any events with three causes (speaking unrestrictedly), they are extremely rare. But the common view comprises also a second claim; it does not just say that in some contexts some causes are ignored. It also says that ignoring them "turns them into" background conditions, and that all background conditions are really just causes (unrestrictedly speaking). In other words, the common view is not just the harmless observation that "cause," like so many other words and phrases, is context-sensitive.

One reason why I don't like the common view is that I can't seem to get myself into a context in which "The presence of oxygen was a cause of the lighting of the match" is true; if the common view is right, I should be able to. But I don't put much weight on this failure, for reasons I will explain in the next section. (I *can* get myself into a context in which "The match lit because oxygen was present" is used to say something true; but, as I will explain, that does not entail that the presence of oxygen was a cause.) Another reason I don't like the common view is that it has the wrong form. The relation that a cause bears to one of its effects is different from the relation a background condition bears to that effect. *Cause* is binary, but *background condition* is ternary: a background condition is not a background condition to an effect alone, it is a background condition to *a cause's causing that effect*. The presence of oxygen is a background

J. L. Mackie (1980, 34–6) (already mentioned), David Lewis (1986a, 162), Fred Dretske—sometimes, anyway (1988, 24)—and Michael Strevens (2008, §6.1). The common view also seems implicit in (Hitchcock and Woodward 2003). These philosophers may not all accept the specific claim about the way in which "cause" is context-sensitive that I attribute to the common view; but none of my arguments depend on that claim. I will have more to say on Dretske's views about background conditions below. (Mackie does not always explain our unwillingness to say "X is a cause of E" by saying that X is excluded from the extension of "cause" in our context; sometimes he says instead that while it is in the extension, and so that the sentence is true, it would be misleading to assert it.)

condition to the striking's causing the lighting, not to the lighting in isolation. But if the common view is true, the background-condition relation is the causal relation, and so has the same arity.

The "wrong arity" objection feeds in to another objection. If *background condition* is binary, and a background condition B to E is, speaking urestrictedly, a cause of E, then B will be a background condition to too many other things as well. Take any event F that E itself causes; since B is a cause E, B will also be a cause of F (unless we're in one of those strange circumstances in which, some allege, causation fails to be transitive). Presumably it follows then that B is a background condition to F,[2] which it surely need not be. If I strike a match, which lights, and the lighting startles you, then the presence of oxygen is a background condition to the lighting, but not to the startling.

One may feel driven to the common view if one cannot see how causes and background conditions could differ metaphysically (rather than pragmatically). But I find it quite easy, at least as a first stab, to articulate a metaphysical difference between causes and background conditions. Causes are *events*; background conditions are not, instead they are *states*. This is, however, just a first stab, and it immediately raises a lot of questions, including: what is the difference between events and states? And: even if background conditions are states, not every state is a background condition—so what does it take for a state to be a background condition?

II

My answers to these questions will constitute my theory of background conditions. Now you won't be surprised when I say that I think

[2] Whether this follows depends on whether there is one single condition that, for all E, excludes things from the extension of "cause of E" in a given context, or whether the condition can depend on E. The statements of the common view I've read don't take a stand on this. It does strike me as more psychologically plausible to think there is one single condition linked to the word "cause," rather than a different condition linked to the multitude of phrases of the form "cause of E."

my theory is pretty good. But my view about why it is good—that you might not expect.

I *don't* think it's good because it's supported by a mass of data. There is, in fact, in my view, very little data to go on. This marks a difference between theories of background conditions and theories of causation. Let me explain.

The core data for testing a theory of causation is made up of particular facts about what did, or didn't, cause what: facts of the form "in scenario X, C was (/was not) a cause of E." We test a theory against this data by seeing if the theory agrees that in X, C was (/not) a cause. Now a particular fact like this is only part of *our* evidence for, or against, a theory of causation if, first, we know that fact to obtain and, second, this knowledge is not theoretically tainted, where the knowledge is theoretically tainted if we used a theory of causation to come to know the fact. (If our knowledge of the fact is based on the theory, we can't also base our knowledge of the theory on the fact!) Luckily there is another way to know whether C was a cause of E that does not use a theory, so it is possible for this kind of knowledge to be part of our evidence. The other way is a method sometimes called "checking your intuitions." In short, to use this method you contemplate the scenario (given to you in a way that leaves open whether C was a cause of E), and see if the claim that C was a cause of E does, or does not, strike you as correct. If it does, and you come to believe it, and this belief constitues knowledge (and a lot of philosophical practice presupposes that beliefs formed using this method can constitute knowledge), then that claim is among your evidence for or against theories of causation.

Now in the case of causation, we have come to know a lot of facts about what caused what using this method. A wide variety of scenarios are such that a causal claim about them strikes us as true, or as false. I say "us": there is widespread agreement. Moreover, people's judgments are relatively strong.

The situation with background conditions is different. I think that we have fewer pre-theoretic opinions about what the background conditions to a given case of causation are, and I think that what opinions we do have are weaker. In fact I suspect that non-philosophers have almost no pre-theoretic opinions about background conditions—and not because they have asked themselves whether, in a given scenario, something is a background condition, and found that they did not know, but because they have never asked themselves questions like that in the first place. "Background condition" just isn't in their vocabulary. I think that if you asked them "Look, that guy struck a match, causing it to light; what do you think are the background conditions to this instance of causation?" they will look at you blankly, or say "what do you mean, background conditions?" (Some evidence: in a search of the world wide web for the term "background condition" only one of the first twenty results—the sixteenth in fact—had anything to do with the cause/background condition contrast.[3])

True, it is not hard to find philosophers making statements like this well-known statement by Hart and Honoré:

> The cause of a great famine in India may be identified by an Indian peasant as the drought, but the World Food Authority may identify the Indian Government's failure to build up food reserves as the cause and the drought as a mere condition. (1985, 35–6)

Hart and Honoré make two claims here. First they claim that a peasant and the World Food Authority may disagree about what "the" cause of the famine was. I think they're definitely right about that. Events typically have many, many causes, but we can get into in contexts in which "cause" is so restricted that it has only one thing in its extension. People who (appear to) disagree about what "the" cause

[3] I used the Duck Duck Go search engine, on September 9th, 2017. The sixteenth result was this story <https://www.yahoo.com/news/macrons-shocking-win-inspires-americans-plotting-independent-2020-bid-190249814.html>, in which "major disillusionment with the political system" is said to be a background condition to Emmanuel Macron's winning of the 2017 French presidential election. The twenty-first result was a philosophical encyclopedia article on causation. After it, the search results continued to have little or nothing to do with the cause/background condition distinction for quite a while.

is are in contexts like that. (If their contexts restrict "cause" in the same way, their disagreement is real; if their contexts restrict "cause" in different ways, they are talking past one another.) These claims depend only on the claim that "cause" is context-sensitive, a claim that I do not dispute. But Hart and Honoré also claim that the World Food Authority "may" assert that the drought was a background condition to the Goverment's causing the famine. I just don't think this likely. In fact, I think that if someone, upon hearing the World Food Authority say that the cause of the famine was the Government's failure to build up food reserves, responded by asking the Authority "but what about the drought? wasn't it also a cause?" the Authority would not respond with "no no, that was just a background condition," but with "yes, okay, that was a cause too, but not the kind of cause we're looking for."

I've suggested that non-philosophers have no opinions about whether something is a background condition. The same isn't true of philosophers; some philosophers are capable of having strong opinions about whether something is a background condition. So there are some opinions out there. But I doubt that many of their opinions are really "pre-theoretic."

The result is that the range of cases that a theory of background conditions has to get right is very small. These are the central, paradigm cases of background conditions—the presence of oxygen being a canonical paradigm case. Now because there are so few such cases, it is not hard for a theory of background conditions to get them right. How are we to decide between those theories? Do we have to throw up our hands and accept that they're all doing equally well?

No, we don't. Fit with the data is not the only good-making feature a theory may have; another is that of "carving nature at its joints." So we should ask of the theories of background conditions that remain how well they do this. Is the way it divides causes from background conditions a very natural way to divide them? A "yes" answer is a mark in favor of a theory. The theory I will propose gets a yes, and that is the main reason I have to offer for accepting it.

While I think it is a nice thing in itself to have a theory of the cause/background condition distinction, I also think the theory I will propose can do useful philosophical work. One place it can do work is in answering some philosophical questions about dispositions; I will discuss those questions in sections X and XI.

III

One part of my thesis, that causes and effects are always events, and so that every cause is an event, is relatively uncontroversial. This claim is a presupposition of a lot of theories of causation, including the most influential theory of the last forty years, namely David Lewis's (Lewis 1986a). Not everyone agrees; there are those who say that causes are facts rather than events, for example D. H. Mellor (Mellor 2004). I reject their view (and will argue against it in chapter 5).[4]

More controversial than the claim that causes are events is the claim that no background condition is an event, that background conditions are instead states. There are theories of events—and here again David Lewis's is an example (Lewis 1986b), as is Jaegwon Kim's influential theory (Kim 1993)—that entail that there is an event that consists in oxygen's being present in this room right now. So if the presence of oxygen is a background condition to the lighting of the match, on these theories of events this background condition is an event, not a state. I think these theories of events are false.

But I am getting a little bit ahead of myself. You can't just say that background conditions are states, add that the presence of oxygen is a state, and have done with it. If you're going to appeal to the event/state distinction to draw the cause/background condition distinction, you need at least a little bit of theory, some principled criterion for when you have an event and when you have a state. Presenting such a criterion is the next thing I want to do.

[4] I have thus changed my mind since writing (Skow 2016); that book came out in favor of fact causation.

The distinction between events and states is commonly used in semantics, in particular in so-called "event semantics" or "Neo-Davidsonian semantics" for natural language. I think their criteria for telling when you have an event and when you have a state are good ones.[5] I will adopt those criteria.

The Neo-Davidsonian program hypothesizes that certain verbs are, despite surface appearances, predicates of events, rather than predicates of the denotations of their arguments. For example, in the sentence "John hit the wall" the verb "hit" has two arguments, "John" (the subject) and "the wall" (the direct object). In logic 101 we would treat "hit" as a two-place predicate, and say that the sentence "John hit the wall" is true iff this predicate is true of the pair ⟨John, the wall⟩. Neo-Davidsonians, by contrast, say that "John hit the wall" has the following truth-condition: there is (in the past) an event E such that E is a hitting, John is the agent of E, and the wall is the object (or "theme") of E. In this truth-condition the verb "hit" has become the predicate "is a hitting," which applies to an event-variable that does not appear in the "surface form" of the original sentence.

Neo-Davidsonians, again, say that "certain" verbs are predicates of events; they don't say that all of them are. So which verbs are (on this theory) predicates of events? The "non-stative" verbs. Non-stative verbs can be distinguished from stative verbs using a variety of tests; one common test uses the notion of progressive aspect. A verb phrase is in the progressive aspect when it consists of "be" followed by the present-participle form of a verb: "Brutus was stabbing Caesar" is in the (past) progressive; "Brutus stabbed Caesar" is not. For the most part, if a verb can occur in the progressive aspect, it is non-stative. Another common test, which will be important later, looks to the form "One thing he/she/it/did was..." If you complete this form with a verb phrase and the result is grammatical, then the verb heading that phrase is non-stative, and vice-versa. "Stab" comes out non-stative

[5] A standard reference is (Parsons 1990).

on this test too, while "believe" does not: "One thing he did was stab Caesar" is grammatical, while "One thing John did was believe that snow is white" is not.[6]

Now Neo-Davidsonians say that, in some sense of "means," if V is a non-stative verb phrase, then "X Ved" means that there was an event that was a Ving by X.[7] Since we are doing metaphysics here we don't need to go that far. Here is the theory I want to endorse:

> If V is a non-stative verb phrase, then necessarily, if X Ved, an event occurred in virtue of this fact.

In many cases we can also say something about the event: again, if the verb phrase is "stab," the corresponding event is a stabbing.

I've been discussing events; what about states? Just as Neo-Davidsonians say that non-stative verbs are really predicates of events, they say that stative verbs are really predicates of states. But as before

[6] As I have written the tests, only the second one gives conditions that are necessary and sufficient; the first is stated only as a fairly reliable "positive" test (if a verb satisfies the test's condition, it is probably non-stative; if a verb does not, the test gives no information). Many think that so-called "achievement" verbs are non-stative but cannot occur in the progressive. Szabó (1994) argues that in fact they can, and concludes that every non-stative verb can occur in the progressive. Comrie (1976, 37–9) discusses progressive uses of stative verbs. A list of other tests that linguists use to distinguish stative from non-stative verbs may be found in (Dowdy 1979) and (Parsons 1990). The division of non-stative verb phrases into subclasses, one of which is the class of achievement verb phrases, comes from (Vendler 1967); something like the distinction goes back to Aristotle. (The other two sub-classes are the "activities" and the "accomplishments"; these distinctions won't matter, so I won't discuss them.)

[7] Here "Ving" is the gerundial noun derived from the verb phrase V, as the noun "stabbing" in "there were ten stabbings" is derived from "stab." I noted in chapter 1 that some non-stative verbs lack corresponding gerundial nouns. So there are non-stative verbs for which you cannot get a true instance in English of the schema " 'X Ved' means that there was a Ving by X." I'm using the schema here as just a rough way to get across the Neo-Davidsonian view. (It is open to Neo-Davidsonians to use a meta-language that departs from English by containing a gerundial noun for every (English) non-stative verb. Then the schema does have a true instance for every non-stative verb.)

we can get by with something weaker: if V is a stative verb phrase then, necessarily, if X Ved, a state obtained in virtue of this fact.[8]

So far the theory just gives us sufficient conditions for the existence of events and states; in a minute I'll suggest a way to make them necessary and sufficient.

These criteria classify the striking of the match and the presence of oxygen the right way. "Strike" is non-stative, so the truth of "Jones struck the match" gives us an event, the striking, that is at least eligible (by virtue of being an event) to be a cause of the lighting. But "was present," in "Oxygen was present in the room," is stative, so the truth of that sentence gives us only a state, something not eligible to be a cause. Instead it is eligible to be a background condition. (Presumably "the presence of oxygen" denotes that state.)

Let me agree right away that these criteria hardly give us (metaphysicians) what we most want out of a theory of events and states, namely a catalogue of the essential, defining features of events and

[8] Do we have names for these states in English? We do have phrases like "the state of believing that snow is white," but as this phrase is ordinarily used it denotes the property of believing that snow is white, rather than a state corresponding to a truth of the form "X believes that snow is white." Note that if Jones and Smith both believe that snow is white we can say that they are both in the state of believing that snow is white. But it is the property that they have in common (they both instantiate it), not any state of the kind that a theory of the event/state contrast is about; the theory of states says that "Jones believes" corresponds to a different state than does "Smith believes."

To "Jones believes that snow is white" we do have the corresponding phrase "Jones' believing that snow is white." But this differs in important ways from phrases like "Jones' stabbing of Smith." The "believing" phrase is less "noun-like" in its behavior than the "stabbing" phrase. For example in the latter you can move "Jones" and replace it with a determiner, to get "the stabbing of Smith by Jones," but "the believing that snow is white by Jones" is ungrammatical. Still, some linguists do take "Jones' believing that snow is white" to be a noun phrase, for example Pullum (1991). So it could denote a state. (Pullum doesn't say what it denotes. Vendler (1967, ch5) holds that such "imperfect nominals" denote not states but "fact-like" entities that he calls states of affairs; Bennett (1988) and Steward (1997) agree.)

Mourelatos observes that we refer to states "typically through use not of the gerund but of specially associated deverbative nouns" (1978, 429); "hate" and "love" for example are the deverbative nouns associated with the stative verbs that share their spelling.

states. It would hardly do to say that it is essential to events that one of them exists when you can state a truth using a non-stative verb phrase. So it is natural to ask at this point whether I have anything more metaphysical, less linguistic, to say about events and states. I do . . . though the stative verb/non-stative verb distinction still comes into it.

But before I say more, it's worth asking Neo-Davidsonians if they have views on the metaphysical differences between events and states. I have not found them very helpful here. Parsons takes a stab at listing some differences in his (1990, 20–1). For example he claims that only events can culminate, or finish; states cannot. (To be precise, the claim is that if something that is either an event or state culminated, it is an event; he does not say that every kind of event can culminate.) Now "culminate" is a technical term, and I'm not sure what it means, so the claim about culmination isn't much help. What about the claim about finishing? It's not helpful either. The problem is that Parsons uses "finish" as an intransitive verb. He wants to say that, since a stabbing can finish, stabbings are events. But I can't make sense of "The stabbing finished" when I'm told "finish" is to be understood intransitively. Of course, if I'm allowed to interpret "finish" transitively, then I can make sense of the target sentence by supplying a missing object: certainly a stabbing can finish *happening*. But then it looks like what is distinctive of events is not that they can finish, but that they can happen. And the claim that what sets events apart from states is that only they can happen, while true, isn't very illuminating.

IV

There is a natural picture that many metaphysicians accept according to which different kinds of things exist at different "levels" of reality. There are the things that exist at the fundamental level, on the "metaphysical ground floor"; and then there are things that exist at some less fundamental level. Most metaphysicians who believe in events do not put them on the ground floor. But if they're not on the ground floor, then their existence is determined by, grounded in, facts about other kinds of things. What other kinds of things? Well one suggestion is that the catalogue of which events exist is grounded in the catalogue of what properties are instantiated by what things. But I think this is wrong. It is *states* that are grounded in properties; events are grounded instead in *acts*.

What is an act? Acts, as I am conceiving of them, are in many ways very similar to properties. I take properties to be abstract things that can be instantiated, and which are definitionally linked to corresponding predicates, in the sense that it is definitional of the property of being red that something instantiates it iff that thing is red. Similarly, acts are abstract things that can be *engaged in*, and are definitionally linked to corresponding predicates, in the sense that it is definitional of the act of hitting the wall that something is engaged in it iff that thing is hitting the wall. ("Act" is also sometimes used as a general term for a kind of event, as in "There were three separate acts of sabotage during the last election," which I will say more about below; this is not how I'm using "act" here.)

How then are events grounded in acts, and states in properties? Like this:

> If something is engaged in an act, then in virtue of that fact an event is occurring, and conversely.[9] If something instantiates a property, then a corresponding states obtains, and conversely.[10]

This claim immediately raises the question, when is something engaged in an act, and when does it instead instantiate a property?

[9] Here the quantifiers are to be given wide scope. Disambiguated, the claim looks like this: for any thing X and any act Z, if X is engaged in Z, then in virtue of the fact that X is engaged in Z an event is occurring; and if an event is occurring, then there is an X and a Z such that X is engaged in Z and that event is occurring in virtue of the fact that X is engaged in Z.

[10] This isn't quite right; some acts are engaged in collectively by more than one thing, and the same goes for properties (in that case we call them relations).

This question might prompt a related question, one that is really an objection in disguise: don't things "always" instantiate properties, even when they're (also) engaged in acts, in which case doesn't the theory give us "too many" states? In non-interrogative form the objection can be put like this: to every predicate there corresponds a property, so for any truth "P(X)" there is a truth "X instantiates P," so for every truth there is a state; but the theory is supposed to be discriminating, it's supposed to say that some but only some truths correspond to the obtaining of a state, while other, distinct truths correspond to the occurrence of an event.

This objection relies on what we can call the "naive" theory of the correspondence between predicates and properties. That theory says that for every predicate there is a property; more precisely, it says that if X is a predicate, then there is a property P with the feature that it is definitional of P that something satisfies X iff it instantiates P. This theory, of course, is inconsistent in classical logic, because of Russell's Paradox for properties (the theory entails that the property corresponding to "X does not instantiate itself" both does and does not instantiate itself), and sometimes what is said to be naive about the theory is that it cannot survive the sophistication that comes from exposure to Russell's paradox. But there is another way in which it is naive; it also cannot survive exposure to the distinction between stative and non-stative verbs.

I take it to be part of the theory that if a predicate has a corresponding property, then that property has a canonical name that is generated from materials in the predicate according to a standard recipe. We're all familiar with the recipe: the canonical name for the property corresponding to "X is red" is "the property of being red"; the canonical name for the property corresponding to "X believes that snow is white" is "the property of believing that snow is white." In general, to construct the canonical name you take the main verb of the predicate and put it in its present-participle form, then prefix it with "the property of."[11]

[11] For evidence that this is indeed part of the naive theory of properties here is Benjamin Schnieder characterizing the theory in a paper called "The Naïve

Now the interesting thing about this procedure for generating canonical names is that it does not work for non-stative verbs. If we try to apply it, for example, to "X is striking a match," we get "the property of being striking a match," which is nonsense, as is "the property of striking a match." I conclude from this that there is no property corresponding to the predicate "X is striking a match," or to any predicate "X VP" where "VP" is replaced by a non-stative verb phrase.[12]

Even if it's true that properties go only with stative verb phrases, there certainly is a property which is such that, necessarily, someone instantiates it iff he is striking a match. It is the property of being someone who is striking a match. (Necessarily, someone has this property iff he is someone who is striking a match, and necessarily, someone is someone who is striking a match iff he is striking a match.) Why can't this property be the property that corresponds to the predicate "X is striking a match"? True, "the property of being someone who is striking a match," while canonical in form, is not derived from "X is striking a match" (it is of course derived instead from "X is someone who is striking a match"). But how bad would it be, really, to reject the thesis that the property corresponding to "X is striking a match," if it exists, has a canonical name generated in the usual way?

If the only alternative were to say that there is nothing that stands to "X is striking a match" as the property of being red stands to "X is red," I might agree that this is not so bad. But on my view there is

Conception of Properties" (2017, 326): "A canonical way to refer to properties is by the use of nominalizations ('the property of being wise') derived from predicates (the verb phrase 'be wise')."

[12] Indeed, it is interesting that Schnieder in (Schnieder 2017, 326) states the naive theory of properties at a level of generality that excludes non-stative verbs. True, he starts by saying that according to the theory "Possession or exemplification of properties is intimately tied to true predication," which suggests that if "Jones hit the wall" is true, then Jones has some corresponding property. But when he gives his general statement of the theory, he says that it consists of the inference rule "x is F/therefore x has the property of being F" and its converse. And predications of the form "x is F" are stative. So this rule does not entail that there are properties corresponding to non-stative predicates.

something, namely the act of striking a match. In general, I want to endorse a sophisticated theory of properties that says that to every *stative* predicate there corresponds a property (and that properties correspond only to stative predicates); that leaves room for a theory of acts that says that to every non-stative predicate there corresponds an act:

- Properties: for any stative predicate S there is a corresponding property P which has the feature that it is definitional of P that something satisfies S iff it instantiates P.
- Acts: for any non-stative predicate N there is a corresponding act A with the feature that it is definitional of A that something satisfies N iff it is engaged in A.[13]

I am striking a match, for example, if and only if I am engaged in the act of striking a match.

As evidence for this theory I have cited the fact that the standard method for getting a property-name from a predicate only works on statives. Further evidence is the fact that there is a standard method for getting an act-name from a predicate, and it only works on non-statives. From "X is striking a match" we can get "the act of striking a match," but if the method is applied to "X is red" we get "the act of being red," which names nothing.

It makes sense that the things that correspond to non-stative verbs are acts when we look back at one of the tests for non-stative verbs that I gave earlier. The test said that only non-stative verbs make sense in the form "One thing he/she/it did was…" So if you did something, there is an answer to the question of what it is you did, an answer that

[13] I am assuming that N ranges over predicates the main verb of which is in the present tense. Also, this theory is restricted to one-place non-stative predicates in the active voice: so "X is hitting the ball" is a possible value for N while "X is being hit by a ball" and "X is hitting Y" are not. This will matter in chapter 5. In that chapter I will also in effect make a further restriction: that N not include in its range predicates of the form "X is failing to ϕ" or predicates like it, where predicates like it are those in which the main verb takes a "raised subject"; see that chapter for the meaning of this restriction.

(if expressible in English) uses a non-stative verb phrase. But it is also true that if you did something you engaged in an act. Presumably the act you engaged in corresponds to the non-stative verb phrase that describes what you did. (In fact, Anthony Kenny, in one of the first significant discussions of the stative/non-stative distinction, called non-stative verbs "verbs of action" (2003 [1963], 108, 121).)

The view I am defending is more liberal about what acts there are and which things are engaged in them than some others—or at least than those others seem to be. If a proton is accelerating, then, since "accelerate" is a non-stative verb, there is an act, the act of accelerating, and the proton is engaged in that act. You might object to this that protons aren't agents, they can't act. But this would be a mistake. What is true is that protons can't act intentionally or act for reasons. But acting and acting for reasons (or acting intentionally) are different things. And as I am using "act," it is sufficient for acting that one is doing something, and all it takes for one to be doing something is for something of the form "one thing one is doing is..." to be true. The proton meets the condition, for one thing the proton is doing is accelerating. There is a tradition in the theory of action of using "act" to mean "act done for reasons"; on this usage, yes, no proton ever acts. But to assert this is not to reject my thesis.[14]

[14] Here I draw on (Setiya 2009). Setiya also asserts that acts go with non-stative verbs (Setiya 2009, 2013). Although Setiya's work was the first place I came across this view, it is not the first place the view appears. I was surprised to find it advanced already by Judith Thomson in 1977. It's all there in her book: the idea that acts (she uses "activities") go with, and only with, non-stative verbs, and that whenever something engages in an act, there is a corresponding event (Thomson 1977, 114–24). She does not, however, use the theory to do any work: although she writes that "I rather like this account of events," she goes on to say that "nothing important in what follows will turn on accepting" it (123, 124).

Thomson says she developed these ideas from some remarks by Nicholas Wolterstorff, so I went and read his book. And back in 1970 *he's* already got the distinction between properties and acts, and the claims that properties go with stative verbs, that acts go with non-stative verbs, and that acts are things that may be done (1970, 72–7). It is a shame that Wolterstorff's and Thomson's ideas are not more widely known.

Sometimes "act" is used both as a general term for a kind of event (rather than a kind of abstract object), and is restricted to intentional acts; this is what is going on when Davidson asserts that "a man is the agent of an act if what he does can be described under an aspect that makes it intentional" (2001a, 46). If a man is turning his hand intentionally, and by turning his hand is spilling his coffee, but is not spilling his coffee intentionally, then on Davidson's view there is just one event, which is both a turning of the man's hand and a spilling of the man's coffee; Davidson must face the question of whether this event is, or is not, an act, in the sense he is using "act." Davidson's answer is that the event is an act if and only if there is at least one description of the event on which it is a doing of something intentionally (so in this case, the event is an act). I do not mean by "act" what Davidson does when he asserts his thesis.

V

Again, the theory of events linked to my theory of acts says

> If something is engaged in an act, then in virtue of this fact an event is occurring, and conversely. If something instantiates a property, then in virtue of this fact a state obtains, and conversely.

Jaegwon Kim popularized the theory that events (not states) go with properties (not with acts). The "existence condition" of his "property-exemplification account of events" says "Event [*x*, *P*, *t*] exists just in case the substance *x* has the property *P* at time *t*" (1993, 35). If his theory is right, my theory is wrong.

We have to be careful interpreting Kim's theory when comparing it to mine, because Kim was willing to ride roughshod over the distinction between states and events: "There are . . . good reasons for not taking this dichotomy . . . of events and states, too seriously at the initial stage of developing a theory of events" (33). In his theory he uses "events" as a term neutral between what I call events and what I call states. Let's keep that in mind.

Kim's theory is false because it fails to entail the existence of events that it certainly should. Kim at one point addressed the objection that his view cannot be right, the existence of an event cannot consist in something's having a property, since events involve change, but having a property does not involve change. Kim replied: "some properties already imply changes in the substance that has them; for example, fading in color, falling, and freezing" (34). Kim is arguing that his theory does entail the existence of such events as freezings, because the requisite property exists. He tries to identify the property by using the word "freezing." But there is no such property as "the property of freezing" that something has iff it is (in the process of) freezing.[15] Kim's theory says that there is a freezing here where this water is iff this water has "the property of freezing";[16] since there is a freezing here, but the water does not have this property (there is no such property), Kim's theory is false.

While there is no such property as the property of freezing, there is the act of freezing, and water that is freezing is engaged in this act. The counterexample does not work if your theory links the existence of events, not to the having of properties, but to the engaging-in of acts, as mine does.

[15] Here and throughout "X is freezing" is used so that it applies to something that is cooling to below 0 degrees celcius. There is another use of "X is freezing" where "freezing" is not the present-participle of "freeze" but is instead an adjective; on this use, "X is freezing" just means "X's temperature is at or below 0 degrees celcius." The existence of this meaning shows that there is such a property as the property of being freezing; it is just that it is the property had by something iff that thing's temperature is at or below 0. This is not the property Kim had in mind, since only when "freezing" is read as the present-participle of "freeze" would "X is freezing" imply a change in X.

[16] Technically, Kim's theory says that the event [this portion of water, the property of freezing, now] exists (not: a freezing exists) iff this portion of water now has the property of freezing. But presumably the event [this portion of water, the property of freezing, now], if it exists, is a freezing.

A minute ago we saw Kim mention that idea that events involve change. Here's a more precise statement of the idea: whenever something changes, an event occurs in virtue of that fact; and these are the only facts in virtue of which an event occurs. If this theory is true, mine is false. "Stand stock still" is a non-stative verb phrase—"What he did was stand stock still" is grammatical. So my theory entails that when someone stands stock still, an event occurs in virtue of this fact. But standing stock still is not a way of changing. So the "involves change" theory says that it is false that when someone stands stock still, an event occurs in virtue of this fact.[17]

Why prefer my theory to the "involves change" theory? Both get our paradigm example right: striking a match involves change, while it seems that oxygen can be present in a room without anything changing. (I mean, in the actual world the oxygen molecules will be moving around and so changing their positions, but while this is what physics requires for oxygen to be in a room—a room at room temperature anyway—it does not seem to be required by metaphysics.)

I have two reasons for preferring my theory. We have two theories of the kinds of facts in virtue of which events occur. Other things being equal, a theory is better if the line it draws around the kinds of facts in virtue of which events occur marks a deeper "joint in nature." Now I admit that the line between changing and failing to

[17] Tons of people have said that events involve change; one important reference is (Lombard 1986). Steward (1997) also takes standings-still to refute such theories. Others say that to satisfy a non-stative verb phrase something must change (see for example Huddleston 2002, 119), but "stand stock still" shows this view to be false.

Some people who stand stock still change while they are standing. When I stand stock still I also always breathe, and breathing is a way of changing. But breathing is also something that you can do; "breathe" is non-stative. So in fact when I stand still both theories say that an event is happening.

These observations do not show the theories to be equivalent. The theories don't just say when an event is happening, they also make claims about the fact in virtue of which the event is happening. If I always breathe when I stand still, both theories say that at least one event occurs whenever I stand still. But my theory says that an event is occurring in virtue of the fact that I am standing still; the "involves change" theory does not.

change marks a joint in nature of a sort. But I think the line my theory focuses on, the line between engaging in an act and instantiating a property, marks a deeper joint in nature. That makes my theory better.

My theory is also better because, if it is true, events can do the work I want them to do. That work is to be the kind of thing which may cause and be caused. A core claim of my theory of background conditions is that events, and only events, can cause, or be caused. If the "involves change" theory of events is true, then this claim is false. Events cannot do the needed work. The problem is that "non-changes" can be causes. My standing still can cause you to become angry (maybe you have ordered me to move).[18]

My view is not light-years away from the idea that events involve change, and can explain why that idea might seem plausible. For a vast number of non-stative predicates it is true that anything that satisfies them is changing. Also, "change" is itself a non-stative verb. If one generalizes too hastily, one might form the opinion that "change" is the most general non-stative verb, in the sense that all other non-stative verbs entail it; so if some event E occurs, then something satisfies some non-stative predicate N, and so also satisfies "...is changing"—voila. But, as I have explained, this inference is bad.

Where are we? I've proposed a criterion for distinguishing events from states, as part of the larger project of giving a theory of background conditions. Causes are events, I hold, and background

[18] One might ask at this point: "Here you use as a premise the claim that your standing still can cause something. But you reject the claim that oxygen's being in the room can cause something. What justifies this different treatment? Aren't these equally intuitive?" My best answer to these questions appeals to claims I don't defend until chapter 5. Briefly: in that chapter I argue that agent causation is basic. What justifies the different treatment of the two claims is that they are not equally intuitive when translated into the language of agent causation. In that language they become: "Oxygen can cause something by being present" and "I can cause something by standing still." Put this way, I think people will agree that only the second can be true. See chapter 5 for the full argument that "X caused Y by Zing" can only be true if what goes in for "Zing" is non-stative.

conditions are states. Now my criterion says the right thing about the striking/oxygen example, but this is just one example, and we will want to make sure that other canonical examples of the cause/background condition contrast line up with the event/state distinction as I have drawn it. There are controversial cases. For example, it seems to be a platitude among those who have thought about the topic that beliefs and desires cause actions. If I go to the store to get some milk, is not my desire for milk one of the causes of my going? But we can report what someone desires using "want," and "want" is a stative verb: one says "I want milk," not "I am wanting milk." So does this example show my theory to be false?

I want to postpone this question until later in this chapter. For I have barely started laying out my theory, and the time to deal with difficult cases is after I have finished. A theory of background conditions cannot just say that background conditions are states rather than events. For given a cause C and one of its effects E, obviously not every state is a background condition to C's causing E. When I strike the match, not only is it true that the room contains oxygen, it is also true that the room is illuminated. And there are states corresponding to both of these facts. But the state of the room that consists in its being illuminated is not a background condition to the lighting of the match. A theory of background conditions must say something about what makes the difference.

VI

In principle, a theory of what it takes for a state to be a background condition is as difficult to produce as a theory of what it takes for an event to be a cause. Since no one has a completely adequate theory of causation, it would be too much to expect me to have a completely adequate theory of background conditions. And since my theory of background conditions isn't built on decades of chisholming away at an initially plausible idea, it would be too much to expect it to have the level of sophistication that today's theories of causation have. My aim is a theory that is a good start.

One thought you might have is that background conditions are just like causes, except for being states. For example, many philosophers think that causation can be analyzed in counterfactual terms. No counterfactual theory is problem-free, of course, but let us just take as an example the simple counterfactual theory of causation, and set aside the kinds of examples that make trouble for it:

> For C to be a cause of E is for it to be the case that C and E are events, and had C not happened, E would not have happened.

If you think that background conditions are just like causes, except for being states, then you might advance a corresponding counterfactual theory of background conditions:

> For S to be a background condition to C's causing E is for it to be the case that S is a state and C and E are events, and had S not obtained, E would not have happened.

I think this theory is false. Of course I certainly agree that had there been no oxygen in the room, the match would not have lit. What I deny is that the truth of this conditional is what it is for the presence of oxygen to be a background condition.

One objection to the counterfactual theory is that its requirements for S to be a background condition to C's causing E do not mention C at all. But as the way I've written the left-hand side of the theory indicates, and as I argued at the beginning of this chapter, the "background condition" relation has three argument places, not two.

This problem seems easy enough to fix; just change the consequent of the counterfactual:

> For S to be a background condition to C's causing E is for it to be the case that S is a state and C and E are events, and had S not obtained, C would not have been a cause of E.

This second counterfactual theory is better but still not right. The right connective to use on the right-hand side is not the counterfactual connective, but the explanatory connective "X is a reason why Z." Counterfactual dependence is a good test for explanatoriness, but only a good test; there are cases where the two part ways (for example cases of causal overdetermination), and when they do it is explanation, not counterfactuality, that matters to being a background condition. So my initial proposal (which I will refine later) is

> For a state S to be a background condition to C's causing of E is for it to be the case that the fact that S obtains is a reason why C is a cause of E.

Suppose (hypothetically) that a special kind of match will light when struck in the presence of either oxygen or helium. It is struck in the presence of oxygen, and lights. Furthermore, had no oxygen been present, helium would have been present instead. So had no oxygen been present, the striking (still) would have caused the lighting. Then the counterfactual theory wrongly says that the presence of oxygen is not a background condition to the striking's causing the lighting. My theory says the right thing. Even if the striking still would have caused the lighting had oxygen not been present, it remains true that the presence of oxygen is a reason why the striking caused the lighting. I don't have a general theory of explanation from which I can deduce this claim, but it shouldn't be controversial.

A certain event has the effects it does against the background of the background conditions that obtain. In the typical case (when we are not talking about direct causes), a cause C initiates a sequence of events that terminates in a given effect E. The background conditions shape the course of this sequence. They help determine (and thus help explain) what C causes, the "direction" of the causal process that emanates from C, and whether that process terminates in E or some other event. If a canonical metaphor for a cause is one billiard ball rolling along and eventually hitting another, then in the metaphor the background condition is the topography of the table, the bumps and dips that determine where the ball will roll, given how it is moving initially. The initial push on the first ball causes the second ball to begin moving because of the topography of the table, just as my theory says.

Background conditions occupy a different level than causes. The simple counterfactual theory puts causes and background conditions on the same level: just as to be a cause is to be an event that the effect depends on, to be a background condition is to be a state that the effect depends on. If this were true there would be no deep distinction between causes and background conditions. But there is a deep distinction. So in what sense do causes and background conditions occupy different levels on my theory? They occupy different explanatory levels. Causes are reasons why their effects happen. But when C causes E against a background condition S, that S obtains is not, on my view, a reason why E happens. Instead, that S obtains is a reason why C caused E. Instead of pertaining to why E happened, as its causes do, S pertains to the connection between the reasons why E happened (that is, its causes[19]) and E itself. S pertains to why those causes are causes. That's the sense in which background conditions are on a different level.

What I mean by "reason why C is a cause of E" might still not be entirely clear. Is a reason why C happened also a reason why C was a cause of E? You might think the answer is yes, since if C hadn't happened, it certainly wouldn't have caused E. But as I intend this phrase the answer is no. As I intend it to be understood, the reasons why C was a cause of E are the factors that explain why, *given that* C happened, it caused E rather than something else (or nothing at all).[20]

My theory gets the paradigm case of a background condition, namely the presence of oxygen, right; I think it gets any other case that deserves to be called a paradigm case right too (I will discuss

[19] In (Skow 2016) I defended the view that every reason why E happened is either a cause or a ground of E. But I'm here only concerned with reasons that are causes.

[20] Actually, on my view a state cannot be a reason why an event happened. So there's no need to worry that the theory as worded might count a state that is a reason why C happened as a background condition to C's causing E. Still, it is worth making the restriction explicit.

dispositions and mental states as potential counterexamples below). It also, as advertised, makes the distinction between causes and background conditions a natural one, in the metaphysician's sense: the distinction marks a joint in nature because the distinctions used to draw it, the state/event distinction and the reason why E happened/reason why C is a cause of E distinction, mark joints in nature too.

My theory uses the notion of one fact's being a reason why another fact obtains; and this is a philosophically problematic notion. Under what conditions is one fact a reason why another obtained? Philosophers have been debating this question (often under the description "what does it take for one fact to explain another?") at least since Aristotle. Is it a weakness in my theory that it uses a problematic notion? Maybe, but I wish to vehemently oppose the idea that a philosophical theory of a problematic notion, like the notion of a background condition, is useless if the theory makes use of other philosophically problematic notions. At the very least, a theory like that reduces the number of problems you need to solve: once you come up with theories of those other problematic notions, you can use them to produce a theory of background conditions that does not use any problematic notions (and I do happen to have a theory of reasons why; see (Skow 2016)). But I think that even this is too high a bar: showing that two problematic notions are connected, as I say that background conditions and explanation are connected, is progress, even if one does not hope to find theories of either problematic notion in non-problematic terms.

Timothy Williamson provided us with a model. Knowledge is a philosophically problematic notion, but still it can be fruitful in epistemology to start with knowledge, to use it as a basic notion in our theories, and to analyze other notions in terms of it (Williamson 2000). I think that "explanation first" metaphysics and philosophy of science are as promising as "knowledge first" epistemology.[21]

[21] A book-length work of explanation first metaphysics is (Kment 2014).

VII

I do not claim complete novelty for my theory of background conditions, but I've not seen it wholeheartedly endorsed in the form I have endorsed it. Thomasz Bigaj writes that saying that background conditions are states not events is "the simplest and apparently immediate solution" to the problem of distinguishing them from causes, but rejects this claim on the ground that "the underlying distinction between events and non-events... is obviously a vague one," and also on the ground that David Lewis's theory of events counts the presence of oxygen as an event (2005, 601–2). Well probably every distinction is vague, but I don't see any place where the event/state distinction, as I have drawn it, is problematically vague; and the fact that David Lewis's theory of events counts the presence of oxygen as an event is a reason to reject that theory, not to reject this idea about background conditions.

Another example of a philosopher who entertained a theory superficially like mine is Fred Dretske; he wrote that

> background conditions are typically conditions that have persisted, without change, for some time. Hence, the cause, being some *change*, appears as a figure against their (back)ground. (1988, 40)

A "condition that has persisted" sounds like a state; so on one reading Dretske is saying that background conditions are "typically" states. He doesn't go far enough: background conditions are always states. Similarly, he identifies causes with changes. But, again, while causes need to be events, and all changes are events, events need not be changes.

For a third example, Mackie claimed that "we are more ready to say that an event caused a certain effect than that a standing condition did" (1980, 34). But he intended this as an addendum to the common view, not an alternative to it: he is saying that while there are contexts in which events are in the extension of "background condition to E" and states are in the extension of "cause of E" (I take "standing condition" as he uses it to mean state), we tend not to be in such contexts. On my theory there are no such contexts.

Other theories of background conditions do not resemble mine and are, I think, false. I know of two. Hart and Honoré claim that

> [i]n distinguishing between causes and conditions two contrasts are of prime importance. These are the contrasts between what is abnormal and what is normal in relation to any given thing or subject matter, and between a free deliberate action and all other conditions. (1985, 33)

Striking a match is certainly a normal thing in relation to a match, but that does not make the striking a background condition rather than a cause. And there can obviously be causes that are not free deliberate actions. (Admittedly Hart and Honoré were talking about causation *in the law*; and it is plausible that in legal contexts events that are not free deliberate actions are often excluded from the extension of "cause." As I've said, that's not enough to put them in the extension of "background condition" in those contexts.)

The second theory of background conditions is due to Bigaj. After rejecting the idea that background conditions are states, he proposes the following theory (2005, 603):

> B is a background condition for E iff (i) if B hadn't obtained, E wouldn't have happened, and (ii) B holds through a "sphere of similarity" that contains a world in which E does not happen.

Here B and E are events—but events understood liberally, so that what I call states count as events. The theory is proposed on the assumption (which Bigaj knows to be false) that the simple counterfactual theory of causation, that C is a cause of E iff, had C not happened, E wouldn't have happened, is inadequate. Bigaj has the modest aim of proposing a theory of background conditions that is as good as this theory of causation.

Clause (ii) does the heavy lifting. It is what distinguishes causes from background conditions (a cause, Bigaj says, is something that satisfies (i) but not (ii)). If we substitute the definition of "sphere of similarity" into the theory, then it says that if E counterfactually depends on B, then B is a background condition iff there is a world in which E fails to happen, and B still happens/obtains in that world, and also in every world as similar to the actual world as that one. According to this theory, causes and background conditions alike are things such that their failure to occur/obtain is a way for E, the effect, to fail to occur. The difference, according to this theory, is that a background condition has the property that, although its failing to obtain is a way for E to fail to occur, it is not the easiest way.

This theory suffers from a structural flaw that I've mentioned before: a background condition is a background condition to one thing's causing another, and so *background condition* is ternary. But this theory makes it binary, a relation just between the condition and an event.

That's an abstract objection; now for a concrete one. The theory fails because there are circumstances in which the striking of a match counts, on the theory, as a background condition to the lighting, rather than a cause. Suppose for example that it has been raining intermittently all day. Jones, walking outside, has a habit of striking matches for the fun of it, not to see them light, but only because he enjoys the feeling of them scraping against the box. He is so obsessive about striking matches that he takes one out and strikes it every five minutes. At noon Jones takes out a match and strikes it; coincidentally the rain lets up just before noon. Of course the striking causes the match to light. But the striking happens in a "sphere of similarity" that contains a world in which the match does not light. For the nearest world in which the match does not light is one in which the rain persisted while Jones struck the match; the rain soaked the match before Jones finished striking it, and it did not light. If there are any worlds more similar to (or as similar as) the actual world as this one, Jones strikes the match in them too—he's that obsessive. So Bigaj's theory wrongly counts the striking as a background condition, not a cause.[22]

[22] Some might welcome this consequence of the theory: they might say that "background condition" is context-sensitive, and so be happy with the idea that in some contexts a striking of a match can count as a background

VIII

One might try to argue that background conditions are causes after all by exploiting a connection between explanation and causation. If every reason why an event happened is a cause of that event, and background conditions are reasons why events happen, then they are causes. More carefully, the main premise will be

> If one reason why E happened is that R, then the state or event corresponding to the fact that R (as the case may be) is a cause of E.

Adapted to my canonical example, the argument looks like this:

1. One reason why the lighting of the match happened is that the room contained oxygen.
2. If one reason why E happened is that R, then the state/event corresponding to the fact that R was a cause of E.
3. So the state consisting in the room's containing oxygen was a cause of the lighting of the match.

There is room to doubt line 2. Maybe causes are always reasons why their effects happen; but is it really so that every reason why an event happened is a cause of that event? Are there no non-causal explanations? In fact I think there are: there are also "grounding explanations." An example: one reason why the room's temperature increased is that the mean molecular kinetic energy of the molecules

condition. In fact, they might say, Bigaj's theory tells us what the context has to be like for this to happen: it has to be a context where the similarity relation between worlds—which varies from context to context—has it that there is a sphere of similarity throughout which the striking happens that contains a world in which the match fails to light. I, however, think this consequence of Bigaj's theory shows it to be false, since I think there are no contexts in which the striking counts as a background condition. For what it is worth, Bigaj does not think there are contexts like that either, so he would not accept this defense of his theory.

in the room went up. This increase in kinetic energy didn't cause the increase in temperature; it grounded the increase in temperature.[23]

This kind of counterexample isn't much use for resisting the argument, however; line 2 could be replaced by "If one reason why E happened is that R, then the state/event corresponding to the fact that R was a cause *or a ground* of E." The counterexample to line 2 is not a counterexample to this replacement; so we can get a revised conclusion: the state consisting in the room's containing oxygen was a cause or a ground of the lighting. Since it's clearly not a ground, we get the desired conclusion, that it is a cause.

One might try to find counterexamples to the revised premise: maybe there are reasons why an event happened that are neither causes nor grounds. I won't try to do that (not least because I don't think such reasons exist, a thesis I argued for in (Skow 2016)). Instead, my view is that line 1 is false.[24]

My argument that line 1 is false will be defensive. Why think it is true? I think I can explain away any facts that appear to support its truth.

The main strategy for defending line 1 looks like this: in such and such a scenario, someone would be asserting something true by saying "One reason why the lighting of the match happened is that the room contained oxygen" (or the less round about "the match lit because

[23] Many philosophers are under the misapprehension that temperature is average kinetic energy. If they were right it couldn't be that the fact that the temperature went up is grounded in the fact that the average kinetic energy went up, since nothing can ground itself. (There are extra premises in this argument but they seem harmless: if temperature is average kinetic energy then to have a certain temperature is to be such that one's constituents have a certain average kinetic energy; and presumably this latter claim entails that the fact that the temperature went up is identical to the fact that the average kinetic energy went up.) But in fact temperature is not average kinetic energy. One way to be 72 degrees Farenheit is for your constituent molecules to have an average kinetic of such-and-such, but it is not the only way (see Baierlein 1999, 347–9).

[24] In (Skow 2016) I flirted with the idea that line 1 is false, that a background condition to C's causing E is a reason why C caused E but is not a reason why E happened, but did not endorse it; I endorse it now.

oxygen was present," which I'll regard as equivalent here). I agree that there are such scenarios. But there is a gap between *asserting something true by uttering S* and *asserting the proposition that S*. That's the gap I will exploit.

First I recommend changing the example. Obviously in a scenario in which someone offers the words in line 1 as an answer to "why did the match light?" is a scenario in which someone has asked why the match lit. But unless something strange is going on, that's a silly question; we know why matches light. And the silliness is liable to interfere with our judgments about the case. So suppose the match doesn't light when struck, instead it shoots off sparks like a Fourth of July sparkler. That's unusual, and one might very well want to know why. So: Smith is in a room with Jones, Jones picks up a match and strikes it, the match sparkles. Smith asks "Why did the match sparkle?" Jones answers "because it is coated with compound X." Now the match's being coated with compound X is a background condition to the sparkling. It looks like Jones has asserted an analogue of line 1, and that we can conclude that the match's being coated with X helped cause the match to sparkle.

I disagree. Certainly Jones asserted something true; but, in my view, it wasn't the proposition that the match sparkled because it was coated with X. Why not? Well, Jones answered the question Smith asked, but on my view the question Smith asked wasn't the question of why the match sparkled. Yes, that's the question Smith's words expressed, but that needn't be the same as the question Smith asked, and in this case, it is not. On my view, the interrogative sentence Smith uttered is elliptical; Smith left a bit out in the way she worded the question. Made fully explicit, the question she asked is why the match sparkled *as a result of being struck*.

Why think this? Well, think some more about the context of the exchange. Smith saw Jones strike the match, and saw the match sparkle. Certainly Smith knows enough about how matches work, even these strange matches, to know that one cause of the match's sparkling was the striking. I submit that what Smith wanted to know in the situation was not what the other causes of the sparkling were, but instead why the striking resulted in sparkling (rather than, say, lighting). She wanted to know why striking the match had such an unusual effect. Now in an efficient conversation the parties leave out stuff they both already know and can easily fill in themselves. Smith and Jones both know that he struck the match, and that the striking caused the sparkling; after all, they both watched those things happen. So Smith leaves out reference to them. Smith says the words "Why did the match sparkle?" as shorthand for "Why did the match sparkle as a result of being struck?" It follows that when Jones said "because it was coated with compound X" he asserted that the match sparkled as a result of being struck because the match was coated with X. But this proposition is not the (analogue of the) proposition in line 1. Moreover, the truth of this proposition is consistent with (in fact entailed by) my theory.

IX

Fred Dretske, in his book *Explaining Behavior* (1988), claimed that background conditions were causes, just causes of a different kind: structuring causes, he called them. The more "ordinary" kind of cause Dretske called a triggering cause. He introduced the distinction between triggering and structuring causes like this:

> In looking for the cause of a process, we are sometimes looking for the triggering event: what caused the C *which* caused the M. At other times we are looking for the event or events that *shaped* or *structured* the process: what caused C *to* cause M rather than something else. The first type of cause, the triggering cause, causes the process to occur *now*. The second type of cause, the structuring cause, is responsible for its being *this process*, one having M as its product, that occurs now. (42)

Dretske's distinction is between two kinds of causes that a process might have. X is the triggering cause of a process iff X "initiated" the processes. Suppose the process runs from C to M; then X is the triggering cause of the process if X is the cause (in the more usual sense) of C (where the more usual sense is the one in which the object of "cause of..." denotes an event rather than a process). X is a structuring cause of the process, by contrast, if X caused C to cause M (rather than something else). A few pages earlier Dretske had written that a background condition is "the cause of C's causing E" (40). This looks like the same condition as the one defining "structuring cause," so it looks like Dretske identified background conditions with structuring causes.[25]

But this is a bad theory. Background conditions are not structuring causes.

Dretske tries to convince us that background conditions are structuring causes by presenting a battery of examples. Sometimes he just asserts, of a background condition in an example, that it caused C to cause M. That's what happens in his "backspace" example: Dretske observes that the "electrical and mechanical state of the computer" is a background condition to the striking of the backspace key causing the cursor to move (a claim I certainly agree with), and then just asserts that this condition "is the cause of C's causing E" (that is, the cause of the striking's causing the moving) (40), a claim I dispute. The mechanical state of the computer is not a cause of anything, so certainly not a cause of the striking's causing the moving. At least, that's what I think. I'd like Dretske to try to convince me I'm wrong, not just assert the contrary.

A bit more argument peeks out in his next example:

> Since the heat of the oven won't cause the bread to rise unless there is yeast in the dough, we can say that yeast's being in

[25] It later becomes clear that while Dretske thinks that background conditions are structuring causes, he does not accept the converse; he holds that events that cause background conditions to obtain are also structuring causes. This does not matter; the claim I want to evaluate is the one he officially endorses, the weaker claim that background conditions are structuring causes.

> the dough...is the cause of the bread's rising when put in the oven. (40)

For this to be an argument that the yeast's being in the dough (which, let us suppose for the sake of argument, is a background condition to the rising) is a structuring cause, we need a conclusion of the form "the yeast's being in the dough caused C to cause M." We can massage the part after the comma into this form by writing it as "the yeast's being in the dough caused the putting of the bread into the oven to cause the rising of the bread."[26] Now Dretske's argument for this conclusion has one premise, marked by "since": "the heat won't cause the bread to rise unless there is yeast in the dough." This, I think, is an excellent reason to believe that the presence of the yeast is a reason why the putting of the bread into the oven caused the rising of the bread. But the presence of the yeast does not need to cause the putting to cause the rising to be a reason why the putting caused the rising. (And in fact that's my view: it's a reason but not a cause.) So we still don't have a good argument that background conditions are structuring causes.

A similar argument appears in the "thermostat" example:

> it is the present electro-mechanical state of the thermostat—its *being* connected to the furnace, its *being* supplied with electricity, and so on—that constitutes the background conditions *in which* one thing (the movement of the bimetal strip) causes another (furnace ignition).... The thermostat's *being* wired to the furnace, *being* supplied with electricity, and so on is *why* it turns the furnace on when the temperature drops. According to this way of

[26] Another interpretation sticks closer to the surface, but has Dretske making a claim of the wrong form: "the yeast's being in the dough caused the bread's rising (psst: by the way, this event happened after the bread was put in the oven)." I think this is false, but even if true, it wouldn't establish that the yeast's being in the dough was a structuring cause. (Also: in light of Dretske's premise, which I am about to discuss, the example would be better if it focused on the heating of the dough rather than the putting of the bread into the oven. But let's let that slide.)

> looking at things, it is B [this background condition] . . . that is the cause of C's causing E. (40–1)

I agree: the electro-mechanical state of the thermostat is a background condition to the strip's motion's causing the ignition. I also agree: the thermostat's being in that electro-mechanical state is why it turned on the furnace when the temperature dropped.[27] But I don't think this is enough for it to be true that the thermostat's being in that state is the cause of the strip's motion's causing the ignition (or, for that matter, of the temperature drop's causing the ignition).

Dretske seems to be assuming that every reason why the temperature drop caused the furnace ignition is something that caused the temperature drop to cause the ignition. But that's a false assumption.

A structuring cause of the process $C \rightarrow E$, again, is something that caused C to cause E. If I strike a match and thereby cause it to light, and I want to know what the structuring causes were (if there were any), I need to ask what caused the striking to cause the lighting. I'd like to suggest that this is a very weird question. Speaking for myself, but I suspect this is true more generally, the question-form "What caused . . . to cause . . . ?" asks to have its first blank filled with a term for an "ordinary thing" (like a chair, or a hippopotamus) rather than an event, and its second blank filled with a phrase of the form "X to Z," where another term for an ordinary thing goes in for "X." I have no trouble answering "What caused *me* to cause *the match to light*?" (Maybe you did, by ordering me to light the match.) But Dretske is asking a question got by filling in the blanks with terms for events. This just leaves me at sea.

I suspect that Dretske is mixing causal locutions in a way that ensures his claims are false. On the one hand we have the use of "cause" where it is flanked by terms for events, as in "the striking caused the

[27] Dretske actually mentions a why-question about a generic: the question of why the thermostat turns on the furnace when the temperature drops. But the example is better if the why-question targets a specific event.

lighting," "the explosion caused the panic," "the throwing of the rock caused the breaking of the window," and so on. On the other hand we have the use of "cause" in "I caused the match to light by striking it." This clause has a more complicated structure: the subject and direct object are terms for ordinary things, there is a "to" infinitival phrase, describing what the object (match) did, and a "by" phrase, describing what the subject (I) did. These two uses of "cause" seem to me to occupy different levels. On one level, we talk about ordinary things and the things they do; on the other level, we talk about events. If you take a use of "cause" that belongs to the ordinary thing level, the second kind of use, you can "lift" it to the event level and get a sentence using "cause" the first way. From "I caused the match to light by striking it" you can derive "The striking caused the lighting." You can go the other way too, though information has been lost: from "The striking caused the lighting," an event-level causal claim, you can "pull back" to the ordinary thing level to get "Something caused something to light by striking something" (in this case, the second two somethings are the same thing, namely the match).[28]

If this is right, then it would be a kind of confusion to write something in the form "X caused Y to Z," the form that belongs to the ordinary thing level, and put terms for events (or for states, which are on the same level as events) in for "X" and "Y." I suspect that doing this always produces something false. Yet that's what Dretske does, when he asserts that the presence of yeast caused the heating of the bread to cause the rising of the bread.

I started this thought with "I suspect," and maybe my suspicion is wrong. I admit that there are some ways of putting event-names in for "X" and "Y" in "X caused Y to Z" that seem okay: for example, "The striking caused the lighting to happen." I have a backup suspicion that "to happen" is the only thing you can put in for "to Z," when terms for states or events go in for "X" and "Y," to get a truth.

[28] I say more about the relation between these two levels of causal talk in chapter 5.

At the very least, this discussion shows that the argumentative burden on Dretske is heavier than he realizes. "STATE caused EVENT to CAUSE EVENT" is an unusual way of talking, one we're not very familiar with. The right first response to "don't you agree that the presence of yeast caused the heating to cause the rising?" is not the response Dretske seems to expect, that of nodding and saying "well yes that's obviously right." The right first response is to scratch your head, look confused, and not know what to say. In the face of this confusion and uncertainty, we need some kind of argument that background conditions cause events to cause other events. And Dretske hasn't given us one.

I conclude that background conditions are not structuring causes. (I doubt that anything else is a structuring cause either; I suspect that there are no structuring causes.) While Dretske's larger project in *Explaining Behavior* is not my concern here, I will say that I don't think my conclusion is very damaging to that larger project. I suspect that all the work he wanted structuring causes to do could be done by (the things my theory identifies as) background conditions. The idea of something that caused C to cause M, and the idea of a state that explains why C caused M, are not really that far apart. Indeed, my theory of background conditions is not far from what you get if you take Dretske's notion of a structuring cause and remove the requirement that structuring causes be causes.[29]

X

There are several things that philosophers consistently say are causes that are, on my view, not causes but background conditions. These include bases of dispositions, dispositions themselves, beliefs, and desires. I suspect that philosophers say that these things are causes

[29] I say "not far from" because, as I mentioned earlier, Dretske is more liberal with what counts as a structuring cause than I am with what counts as a background condition: events that cause background conditions to obtain are also supposed to be structuring causes.

because they think that these things explain why certain events happen, and then infer from this that they are causes. Anyway, whatever their reason, they would lose nothing by saying that these things are background conditions rather than causes.

To start with dispositions: a common thought about dispositions is that every disposition has a causal basis.[30] What is a causal basis for a disposition? Ignoring a few unimportant qualifications, a causal basis for, say, fragility, is a property B, for example the property of having such-and-such a crystalline structure, with these features:

(i) things that have B and are fragile are fragile because they have B (this is what makes B a basis for fragility);[31]

(ii) For anything X that is fragile because it has B, X's having B would, in David Lewis's words, "join with striking to cause breaking" if X were struck (Lewis 1999, 140)—by which Lewis means that if X were struck and X continued to have B, the striking and X's having B would both be causes of the breaking.

It is clause (ii) that makes having B not just a basis but a causal basis for fragility. So the thesis that every disposition has a causal basis says, of the case of fragility, that everything that is fragile has a property B that meets conditions (i) and (ii).

Here is the problem: according to my theory, a piece of glass's having some property, and so in particular its having a property that is a causal basis of fragility, cannot cause anything, so certainly cannot cause the glass to break; for the having of a property does not make for the occurrence of an event, and only events can cause things.

Can this problem be swept away by replying that philosophers who discuss causal bases of dispositions are using "property" in a wider sense than I am? I, recall, have said that only stative predicates

[30] A standard argument for this thesis that is widely cited is in (Prior et al. 1982).

[31] This clause implicitly restricts attention to possible worlds that have the same laws as ours. In worlds with different laws, there could be things that have the same crystalline structure as this wine glass and are fragile but are not fragile because they have that structure.

correspond to properties. Maybe the "properties" that are causal bases of dispositions are supposed to correspond to non-stative predicates. In fact this reply does not work, as can be seen by looking again at the example. One causal basis of fragility is the property of having such-and-such crystalline structure. But "X has such-and-such crystalline structure" is a stative predicate.

So the thesis that every disposition has a causal basis is inconsistent with my theory of background conditions. But I do not regard this inconsistency as a deep problem. For the thesis has a near cousin that is compatible with my theory; and the near cousin is just as good as the original, for the purposes for which the original was formulated.

The near cousin just says that every disposition has a basis that is a suitable background condition for the disposition's triggering condition to cause its manifestation condition. Sticking with fragility for the sake of having a concrete example, the thesis is that every thing that is fragile has a property B that meets the following conditions:

(i) things that have B and are fragile are fragile because they have B (this condition is unchanged);
(ii*) For anything that is fragile because it has B, its having B would "join with striking to cause breaking"—not in the sense that it would, like the striking, be a cause of the breaking, but in the sense that it would be a background condition to the striking's causing the breaking.

The idea behind the thesis that every disposition has a causal basis is the idea that there are no "bare dispositions": the idea is that whenever something is disposed to M in C, there is some underlying state that causes it to M when the triggering condition C obtains. The above revision of the thesis is just as good for opposing the idea that there are no bare dispositions as the original one. It still requires an underlying state; and while the underlying state is not required to be a cause of the manifestation of the disposition, it is still relevant to the causing of the manifestation, because it is required to be a reason why the triggering condition causes the manifestation.

I just put forward the thesis that when something manifests a disposition, say fragility, its having a basis for fragility is a background condition to the striking's causing the breaking. My claim was that this thesis can do the work that the idea that the basis causes the breaking can do. I do not, however, think that the thesis is true.

The verb phrase "has B," where B is a basis for a given disposition, is stative, so when something has a basis for a disposition, there is a corresponding state, and no corresponding event. Similarly, "is disposed to M in C" is stative, so when something has a disposition, there is a corresponding state, and no corresponding event. Now: between a glass's having a basis for fragility and its being fragile, which should be the background condition? (Surely they're not both background conditions.) This question resembles another, and has a similar answer. Stephen Yablo imagines a bird trained to peck when a light turns red (2008, 230). The light turns a specific shade of red, namely scarlet, and the bird pecks. We have two events: the light's turning red, and the light's turning scarlet. Between the two, which caused the bird to peck? (Surely they both didn't.) The intuitive answer is that it was the light's turning red that caused the bird to peck. This answer is right even if the light's turning scarlet passes the standard counterfactual test for being a cause. Imagine that had the light not turned scarlet, it would have turned aquamarine. Then had the light not turned scarlet, the bird would not have pecked. Still, it was the light's turning red, not its turning scarlet, that was the cause.

This verdict supports the thesis that causes must be "proportional" to their effects. Roughly speaking, when an event C happens in a particular way, as turning scarlet is a particular way of turning red, then the "more general" event (in this case, the turning red) is the cause only if (i) the effect depended on C but not on the way in which C happened, and furthermore (ii) it is false that some other event D happened, that C's happening was a way for D to happen, and (i) is true of D. In more detail:

> C is a cause of E only if (ia) had C not happened E wouldn't have happened;[32] (ib) for each way in which C can happen, E would still have happened if C had happened in that way; and (ii) at least one of these is false for any D that happened that had the property that C's happening was a way for D to happen.[33]

In the bird case: (ia) if the light hadn't turned red, the bird wouldn't have pecked; and (ib) for each shade of red, the bird still would have pecked if the light had turned that shade of red.[34] As for (ii), there are many such events D, so let's just check one: another event that happened was the light's turning some color or other, and the light's turning red was a way for it to turn some color or other. But (ib) is false of this D (one way for it to turn a color is for it to turn blue, but it's false that the bird still would have pecked if the light had turned blue). So if either the light's turning red or the light's turning scarlet is a cause of the pecking, it is the light's turning red that is the cause.

I hold that a proportionality requirement also holds for the reasons why a fact obtains. Since B is a background condition to C's causing E only if it is a reason why C caused E, it follows that a proportionality requirement holds for background conditions. Applied to reasons the requirement is this:

> X is a reason why Z only if (ia) had X not obtained, Z would not have; (ib) for each way for X to obtain, Z still would have obtained if X had obtained in that way; and (ii) at least one of these is false for any fact Y that obtained and had the property that X's obtaining was a way for Y to obtain.

[32] A cause will fail to satisfy (ia) if there are backups that would have caused the effect had the cause not occurred. The proportionality criteria are meant to be used in cases where there is no preemption. The same restriction applies to the proportionality criteria for reasons that I will state shortly.

[33] These are not the conditions as Yablo writes them. I've made some modifications partly based on (Weslake 2013).

[34] For this to work we must read the conditional "The bird still would have pecked if the light had turned scarlet" so that it is true if the light actually turns scarlet and the bird pecks.

It follows that if a fragile glass is struck and then manifests fragility, and either the glass's being fragile or the glass's having such-and-such crystalline structure is a reason why the striking caused the breaking (a plausible claim that I take it needs no defense), it is the glass's being fragile that is the reason. So also it is the glass's being fragile that is the background condition to the striking's causing the breaking. For: (ia) if the glass hadn't been fragile, the striking would not have caused the breaking; and (ib) for each way of being fragile (each basis for fragility is a way of being fragile), the striking still would have caused the breaking, if the glass had been fragile in that way. And (ii) is true: there is no F such that being fragile is a way of being F, and (ia) and (ib) are true of being F. In general, I conclude that X's having a disposition D is a background condition to a triggering event for D's causing X to exhibit the manifestation behavior for D. It is a reason why the trigger caused the manifestation.[35]

Since on my view causes and effects must be events, this claim presupposes that the triggering condition and the manifestation condition for every disposition concerns the occurrence of events. I think it uncontroversial that triggering conditions are always events, and I argue that manifestations conditions always are in chapter 3.

XI

Maybe the main question about dispositions that metaphysicians have tried to answer is, what is it to have a certain disposition? What does it take, for example, for something to be fragile? Some have suggested that it is enough for it to satisfy the conditional *if it were struck, it would*

[35] This argument is a replay of an argument that it is having a disposition, not having a basis for that disposition, that causes the manifestations of that disposition. Fara and Choi (2014, section 6) argue for this conclusion using a proportionality requirement. See (Prior et al. 1982) for the standard argument that it is the basis of a disposition, not the disposition itself, that causes its manifestations. Lombard (1986, 73) also suggests that a disposition merely enables the triggering to cause the manifestation, instead of itself being a cause of the manifestation.

break; others have raised problems for this suggestion, and suggested different answers.[36]

Another question about dispositions is, I think, just as important, but much less discussed. We need to ask not just what it is to have a disposition, but also what it is to manifest a disposition. If background conditions are what I say they are, then we can use the concept of a background condition to answer this question.

What am I talking about when I say "manifest a disposition"? I referred to the "manifestation behavior" of dispositions all through the last section, so it is past time to introduce this concept in more detail. Every disposition is at least partly defined by its associated manifestation behavior. That's the behavior something exhibits when it manifests the disposition: there is some behavior (or some class of behaviors) such that whenever something manifests D it does so by exhibiting that behavior (or one from the class). A fragile glass manifests fragility by breaking. An irascible person manifests irascibility by getting angry.

Now obviously exhibiting the manifestation behavior for a disposition is not sufficient for manifesting that disposition. Something can break without manifesting fragility. Take a giant, super-thick piece of iron; it is not fragile. But if Thor were to smite it with his hammer, it would break. In breaking, it would not manifest fragility.[37] The reason why this is so is obvious: you can't manifest a disposition you don't have.

So, again, what is it to manifest a disposition? You might think this question hasn't received much attention because it is so easy to answer. Something manifests a disposition D, you might say, iff it has D, and exhibits the manifestation behavior for D. A glass manifests fragility iff it is fragile and breaks. This is the answer the Thor example most naturally suggests. But in fact this answer is wrong. It is a fascinating

[36] A good overview is (Fara and Choi 2014).

[37] "Fragile" is context-sensitive. In a context in which Thor is talking to a bunch of other gods, the sentence "Everything is fragile!" might be true. In a context like that, "In breaking, the piece of iron manifested fragility" might also be true. The paragraph to which this note is appended is not in a context like that.

and puzzling fact about dispositions that even something that has D can exhibit the manifestation behavior for D without manifesting D. A fragile glass can break without manifesting fragility. An example of how this can happen may be found in the history of a different debate, about the nature of intentional action.

In 1973 Donald Davidson made famous the problem of "deviant causal chains" for causal theories of intentional action. Causal theories say that an action is intentional if it had the right sorts of causes; Davidson specifically considers the theory that

> X Zed intentionally iff X had a belief and a desire that "rationalized" Zing, and that belief and desire caused X to Z.

Davidson then described his famous hiker example: a hiker who wants to be rid of the weight of a companion tied to him by a rope, believes that by letting go of the rope he can be rid of his companion, and is caused to become nervous by this belief and desire; he then drops the rope out of nervousness. The belief and desire rationalize letting go and cause him to let go, but the hiker does not let go intentionally (Davidson 2001c [1973], 79). To solve the problem of deviant causal chains is to find a condition that can be added to the theory's right-hand side to block this kind of counterexample.[38]

Davidson, in presenting his example, meant to refute David Armstrong's causal theory of intentional action, and Armstrong, in his response, observed that the problem of deviant causal chains appears outside the realm of intentional action. Armstrong noted that with respect to a disposition, there can be deviant causal chains from the triggering event to the manifestation behavior of that disposition (1975, 5). If someone strikes a piece of fragile glass, and the glass breaks, and the striking is a cause of the breaking, it does not follow that in breaking the glass manifested fragility. Kieran Setiya

[38] The theory Davidson discusses speaks of beliefs and desires causing X to Z, which they cannot, on my theory, since they are states; since I discuss this problem in the next section, I will pass over it here.

imagines a glass "attached to an explosive device, triggered by contact, that would destroy even a sturdy object" (2016, 17). Strike the glass; the striking causes an explosion which causes the glass to break, but in breaking the glass does not manifest fragility. The same claim is made by John Hyman:

> every disposition can be connected to the kind of occurrence that normally manifests it by a "deviant" or "freakish" causal chain. Take Molière's example of a disposition, the *virtus dormitiva*. A man might take a soporific drug before driving, and the drowsiness induced by the drug might make him crash the car and knock himself unconscious. If this happened, he would lose consciousness because he took the drug, but the exercise of its *virtus dormitiva* would be pre-empted by the crash. (It is true that the drug's soporific power is manifested to some degree in this case, but if that is thought to be a defect in the example we can instead imagine the man taking the drug while driving, choking on it, crashing, etc.) (2014, 116)

For the reasons Hyman gives in his parenthetical remark I will focus on the version of the case where the man chokes on the pill. One of Hyman's points in claiming that there are deviant causal chains to the manifesting of a disposition is that the problem of deviant causal chains is not a special problem for causal theories of intentional action. Another of his points is that the problem of deviant causal chains for causal theories of intentional action is just a special case of the problem of deviant causal chains for dispositions. For according to Hyman intentional actions are manifestations of dispositions: he holds that desires are dispositions and that people manifest those dispositions by acting intentionally.

Hyman also seems to think that the problem for dispositions cannot be solved; at least, he writes that "the exercise of a disposition cannot be defined as a specific kind of sequence or concatenation of events" (116). Others have reached a similar conclusion: Alexander Bird, writing four years earlier, responds to examples like Hyman's by asserting that "the relation 'Mx is the manifestation of disposition Dx in response to stimulus Sx' is ontologically basic and is not reducible" (2010, 164). Even if this is right, a theory like Hyman's does in a sense solve the problem of deviant causal chains for causal theories of intentional action. Hyman, again, uses the notion of manifesting a disposition in his theory: he holds that X ϕed intentionally iff in ϕing X manifested a disposition that is a desire. Davidson's hiker is not a counterexample to this theory. The hiker's desire may have caused him to drop the rope, but since the causal chain is deviant, in dropping the rope the hiker did not manifest the disposition that is his desire. Of course the question of what it is to manifest a disposition remains unanswered. Action theorists might want to pass the buck to their colleagues down the hall. Let the metaphysicians figure out what manifesting a disposition requires! Whether or not the notion of manifesting a disposition can be explained just in terms of event causation, manifesting a disposition is a causal notion, so a theory of intentional action that makes use of it is still a causal theory.

The same strategy appears in Ralph Wedgwood's theory of reasoning (2007, 184–7). Suppose you believe that it is raining and that if it is raining you should take an umbrella. You now form the belief that you should take an umbrella. Under what conditions have you reasoned from the first two beliefs to the third? Wedgwood notes that it is not enough to say that the first two beliefs caused you to form the third; they could cause this via a deviant causal chain. (You strongly associate rain and umbrellas with your first love, so your beliefs cause you to daydream about him; you wishfully imagine that taking an umbrella might make it likely you will run into him, so you come to believe that you should do so, even though you are no longer attending to the weather, and in fact have by now forgotten that it is raining.) Wedgwood's solution is to speak of manifesting a disposition: you reasoned from the first two beliefs to the third iff in forming the third you manifested a disposition to believe propositions rationalized by other beliefs, and this disposition was triggered by the first two beliefs.

As I said, the question of what it is to manifest a disposition has received less attention from metaphysicians than the question of what it is to have a disposition. But the first question is hiding just around the corner in some discussions of the second. The problem of masks is a standard problem for theories of dispositions: something's disposition is masked iff it would fail to manifest the disposition, even if a triggering event occurred. A glass wrapped in bubble-wrap is fragile but would not break if dropped; its fragility is masked. There is also the problem of reverse-masks, or mimics: something mimics a disposition to M in C iff it lacks the disposition, but still would M in C. Superglass is not fragile, but a certain piece of superglass might be monitored by a demon who will destroy it with a lightning bolt if it is dropped. That piece of superglass then mimics fragility. And there it is: the notion of mimicking a disposition contains the seed of the notion of a deviant causal chain for a disposition. To construct a deviant causal chain, a case where something has a disposition and exhibits the manifestation behavior for that disposition without manifesting it, just construct a case where something mimics a disposition it actually has.[39]

If the question of whether the problem of deviant causal chains for dispositions can be solved has been left to the metaphysicians, then—here I am. Let me try my hand at it.

[39] Masks and mimics were so-named by Johnston (1992). The first case of mimicking that I know of is in (Smith 1977, 444). Mimics are counterexamples to the conditional analysis of dispositions: the analysis says that X is disposed to M in C iff, had C obtained, X would have Med, but if X mimics the disposition, the counterfactual is true even though it lacks the disposition. Prior, in her discussion of Smith's example (Prior 1985, 9), tries to solve the problem of mimics by amending the analysis to say, in effect, that X is disposed to M in C iff, had C obtained, C's obtaining would have caused X to M "in the right sort of way"—the same placeholder language some causal theorists of intentional action use to rule out deviant causal chains (they say that an action is intentional when it is caused by a belief and desire in the right sort of way).

If it is true that there is no solution to the problem of deviant causal chains for dispositions, the examples that we have seen so far do not show it. Armstrong knew what to say about those examples back in 1975. He said that when X has a disposition D, and a triggering event for D happens, and this causes X to exhibit the manifestation behavior for D "deviantly," so X does not manifest D; this is because the triggering event causes the manifestation behavior "without the causal co-operation of the state which is the disposition" (1975, 5). Armstrong seems to be assuming that states generally, and dispositions in particular, can be causes, but his proposal does not essentially depend on this assumption. On my view, the way dispositions cooperate in the causing of their manifestation behaviors is by being background conditions to the triggering of those behaviors. And I spell this out by saying that the fact that X had the disposition is a reason why the trigger caused the manifestation. Translated into this framework, Armstrong's claim is that

(0) If X has a disposition D and a triggering event S for D happens and X then Ms (where Ming is the manifestation behavior for D), then in Ming X manifested D only if the state of X's having D was a reason why S caused X's Ming.

The condition in (0) is just the condition that X's having D be a background condition to S's causing X's Ming; that is, (0) is equivalent, given my theory of background conditions, to

(1) If X has a disposition D and a triggering event S for D happens and X then Ms (where Ming is the manifestation behavior for D), then in Ming X manifested D only if the state of X's having D was a background condition to S's causing X's Ming.

Imagine that Setiya's piece of glass was struck, and broke. The striking caused the breaking, but it's clearly false that the glass's being fragile was a reason why the striking caused the breaking. After all, the explosion that broke the glass was strong enough to break even a "sturdy object." Similarly, when Hyman's driver took the sleeping pill, and choked on it, and the choking caused him to crash and the crash caused him to become unconscious, clearly the fact that the pill is disposed to render those who take it unconscious was not a reason why the taking of the pill caused the driver to become unconscious. If all this is right, then both cases fail to satisfy the necessary condition on manifesting a disposition in (1).

Is it really true in both cases that the having of the disposition is not a reason why the trigger caused the manifestation? I think the answer is "obviously, yes" but I do not need to rest my case there. The simple counterfactual theory of causation—C is a cause of E iff, had C not happened, E would not have happened—is false, but this counterfactual is still a useful guide to causation. If some C and E fail to satisfy the counterfactual, that's at least defeasible evidence that C is not a cause of E. The same goes for the simple counterfactual theory of reasons why—that F is a reason why G iff, had it not been that F, it would not have been that G. I do not endorse this theory, but the counterfactual is a useful test for being a reason why.

The dispositions in the examples fail the counterfactual test. Looking just at Hyman's, if the pill had not been disposed to cause unconsciousness, the taking of the pill still would have caused the driver to become unconscious (for it still would have caused him to choke, etc.).

Of course if the pill had never been disposed to cause unconsciousness, perhaps it never would have been in the pill bottle in the man's backpack (it would have been removed through some quality screening process), and so would never have been taken. But this is irrelevant. In assessing whether the pill's being disposed to cause unconsciousness is a reason why the taking of the pill caused the man to become unconscious, we must interpret the counterfactual so that what matters are scenarios in which the pill loses this disposition just before it is taken, not some time long before.

The claim in (1) just gives a necessary condition on manifesting a disposition; a solution to the problem of deviant causal chains requires conditions that are necessary and sufficient. Could the condition in (1) be promoted to one that is both? That is, is the following true?

(2) If X has a disposition D and a triggering event S for D happens and X then Ms (where Ming is the manifestation behavior for D), then in Ming X manifested D if and only if the state of X's having D was a background condition to S's causing X's Ming.

I find it hard to say for sure whether this is right; I do have some doubts. We need to consider the following kind of example:

> A piece of glass is hooked up to a machine. The machine is programmed so that when the glass is struck it quickly checks whether the glass is fragile; if, and only if, the glass is fragile, the machine then breaks the glass (with a blow hard enough to break even shatter-proof glass).

Strike the glass and it breaks; in breaking, it does not manifest its fragility. So: is the glass's being fragile a background condition to the striking's causing the breaking, in this case? If so, we have a counterexample to (2); if not, we do not.

The glass's being fragile is certainly a background condition to one part of the process leading from the striking to the breaking. The machine checks whether the glass is fragile, and if it finds the glass to be fragile, the machine breaks the glass. So the checking (whether the glass is fragile) causes the finding (that the glass is fragile), which causes the breaking; and the fact that the glass is fragile is a background condition to the checking's causing the finding. Now if we have a process C→D→E→F (here, striking→checking→finding→breaking), and B is a background condition to D's causing E, does it follow that B is also a background condition to C's causing F? More generally, if B is a background condition to one link in the chain of causes from X to Y, does it follow that it is also a background condition to X's causing Y? If so, then the glass's being fragile is also a background condition to the striking's causing the breaking, and (2) is false.

I have to admit, though, to being uncertain about whether these claims are true. We can, however, still make progress, because the claims themselves suggest an alternative to (2) to which the example is not a counterexample. First a definition: if we have a process starting with E1 and ending with En (E1 → En for short), and there is a state B with the property that for any two events Ei and Ej in this process for which Ei is a cause of Ej B is a background condition to Ei's causing Ej, then I will say that B is a background condition to the process E1 → En. Here then is my proposal:

(3) If X has a disposition D and a triggering event S for D happens and X then Ms (where Ming is the manifestation behavior for D), then in Ming X manifested D if and only if X's having D was a background condition to the process S → X's Ming.

I (tentatively!) propose this as a solution to the problem of deviant causal chains for dispositions.

I'm proposing (3) against the background of my theory of background conditions. Is my theory required, for (3) to have a chance? Well (3) is unlikely to get off the ground if we plug in the common view of background conditions. The common view says that "background condition" is context-sensitive. That's a problem: although the left-hand side of (3) is also context-sensitive (all dispositional ascriptions are context-sensitive[40]), the context-sensitivities seem unlikely to line up. Also, (3) requires the notion of a background condition to a process; how can the common view define this notion? Defenders of the common view can speak for themselves; I will just point out that it won't do for them to say that something satisfies "background condition to a process C → E" in a context K iff it is a cause of E unrestrictedly speaking but fails to be in the extension of "cause of E" in K. For if they do, the theory will have to count the glass's being fragile, in the souped-up exploding glass example, as a background condition to the striking → breaking process.[41] But that's the wrong answer! It makes the example a counterexample to (3).

[40] This is connected to the fact that dispositional ascriptions are gradable: one thing can be more fragile than another; one thing can be more disposed to break than another (see Manley and Wasserman 2007).

[41] The glass's being fragile is certainly a background condition to the checking → finding process and so, given this defintion of "background condition to a process," a cause of the finding; since the finding is a cause of the

Go back to my claim about what it takes to be a background condition to a process—is it too demanding? The presence of oxygen is supposed to be a background condition to the striking's causing the lighting; surely it is also a background condition to the process striking→ lighting. But this process at least includes as an intermediate event the heating of the match head. The striking causes the heating, which causes the lighting. Now the presence of oxygen is definitely a reason why the heating caused the lighting. But it doesn't seem to be a reason why the striking caused the heating. (The striking still would have caused the heating had it happened in a vacuum.) So my view says that the presence of oxygen is not a background condition to the striking→ lighting process.[42]

Instead of showing my view to be false, I just think this shows we should revise our view about what processes the presence of oxygen is a background condition to. It isn't a background condition to the entire striking→ lighting process, only to the heating→ lighting subprocess. Whatever inclination we have to say that it is a background condition to the larger process I chalk up to our usually not distinguishing the heating as a distinct event in the process.

Kieran Setiya, when discussing the problem of deviant causal chains for causal theories of intentional action, suggested that "the crucial concept is that of *guidance*: when an agent ϕs intentionally, he *wants* to ϕ, and this desire not only causes but continues to guide behavior towards its object" (Setiya 2007, 32). (He takes the notion of guidance from Frankfurt (1988).) If we translate this proposal from the context of intentional action to the context of dispositions, we get this proposal: when X Ms and in Ming manifests a disposition D, X's having D not only causes X to M, but guides X toward Ming. Now I've already said that I don't think that X's having D causes X to M. But set that requirement aside; it is the second clause of the proposal that I am interested in, the clause about guidance. I think that there is an interpretation of "guides X toward Ming" on which this second clause

breaking, the glass's being fragile is as well—causation may not be transitive, but this doesn't seem like the kind of cause where transitivity fails.

[42] Dan Greco raised this objection.

follows from my theory. I said earlier that when one billiard ball rolls toward and hits another, the topography of the table is a background condition to the rolling and hitting. It is natural to say also that the topography of the table guides the first ball as it rolls and then hits the second. I also said earlier that when a cause initiates a process, the background conditions shape the course of this process: they determine what the rest of the process looks like, given how it started. I submit that this is a sense in which background conditions guide the processes to which they are background conditions: a background condition to a process E1→En guides the process from E1 to En. But then the claim that, when a triggering event for D happens, X's having D guides X to exhibit the manifestation behavior for D, follows from my proposal (3).

I have defended, and have been exploring some consequences of, the idea that dispositions are background conditions not causes. Against this idea one might argue that we often cite the fact that something has a disposition to explain why it exhibited that disposition's manifestation behavior. The objection might be put like this: "We often answer the question 'Why did the glass break?' by saying that it was fragile.[43] But if the fact that the glass was fragile is a reason why the glass broke, doesn't it have to be that the fragility was a cause of the breaking?"

I've already discussed an argument essentially the same as this one. I want to say the same thing here. It is consistent with my view that we can answer a question phrased as "Why did X Z?" with "Because X was S," were "was S" is a stative predicate, as long as the initial question is elliptical for "Why did X Z *as a result of*...?," where what goes in for the ellipsis describes a cause of X's Zing. For to ask this is to ask why one event caused another—and to this kind of question, it is exactly right to answer by describing the background conditions that enabled the cause to cause its effect. When "Because it was fragile" is

[43] Actually there has been philosophical controversy over whether you can answer this why-question this way. It has seemed to some too much like saying that the opium put me to sleep because it has the dormative virtue. I think the answer, at least in the case of the glass, is fine, so I do have to face the consequences of admitting that it is fine.

an answer to "Why did the glass break?", the question actually asked is the question of why the glass broke as a result of being struck.

Another objection one might raise focuses on the possibility of dispositions that lack triggering conditions. I've said that whenever something has a certain disposition, and manifests that disposition, the state of its having that disposition is a background condition to the trigger's causing the manifestation. But it has been claimed that some dispositions lack triggering conditions. If that is true, and D is such a disposition, then if something X has D, and manifests D, it cannot be that its having D is a background condition to the trigger's causing the manifestation—for there is supposed to be no trigger in these cases. Since X's having D seems clearly relevant in some way to X's manifesting D, the only option left seems to be to say that X's having D is a cause of its manifesting D—just what I have denied.[44]

Let's look at the case for dispositions that lack triggering conditions.[45] Here are Manley and Wasserman (2008):

> It is commonly assumed that every disposition is associated with a particular set of triggering conditions: *fragility* with

[44] Daniel Nolan (2015) has argued that there are "non-causal" dispositions that nevertheless have triggering conditions; the triggering conditions fail to be causes of the manifestations of those dispositions. Since I don't find his examples plausible, I won't discuss them.

[45] The existence of triggerless dispositions is central to Barbara Vetter's views about dispositions (2015). Three examples she proposes of such dispositions are essentially the same as the ones I am about to discuss. She also suggests that an "electron's disposition…to exert a force which stands in a certain mathematical correlation to surrounding charges and their distance" is triggerless (98). But when an electron exerts a force with a certain magnitude on some other charged body that is some distance D away, and thereby manifests this disposition, what triggers the manifestation is that other particle's coming to be distance D away. In more detail, Coulomb's law says that the electrostatic force a body of charge Q exerts on another body of charge q that is distance r away is equal in magnitude to $k|Qq|/r$; so the relevant disposition that having charge Q confers is the disposition to exert a force of magnitude $k|Qq|/r$ on any body that becomes a body that has charge q and is distance r away. Far from being an example of a triggerless disposition, this seems to me a paradigm case of a disposition that has a trigger; the triggering condition is coming to be distance r away from a body of charge q.

> *being stressed* and so on. We have so far gone along with this assumption, but we submit that it is plausible only given a paltry diet of examples. In fact there are plenty of dispositions that do not have any particular triggering conditions. Suppose someone is highly disposed to talk, but there is no particular kind of situation that elicits this response from him. He is disposed to talk when happy, when sad, with others or by himself—he is just generally loquacious. Or take someone whose frequent bouts of anger are not particularly tied to provocation; surely it is relevant to one's degree of irascibility if one gets angry for no reason. The point is that talking in any situation at all can manifest loquaciousness, and anger in any situation at all can manifest irascibility. Perhaps even fragility is just the disposition to break—after all, breaking for no reason at all would be relevant to a thing's degree of fragility. (72–3)

This is not convincing. Suppose someone is very loquacious, and that on a certain occasion they begin to talk. Could it really be that nothing triggered his talking on that occasion? Surely there are plenty of events that are good candidates to be causes: maybe someone just walked up to him, maybe he just arrived at a party, maybe he just arrived home. The kind of person Manley and Wasserman are imagining seems to me not to be one who frequently begins talking without anything causing him to begin, but one who can be caused to begin to talk by pretty much anything. Far from being an example of a triggerless disposition, this is an example of a disposition that almost anything can trigger.

Manley and Wasserman claim that a loquacious person could frequently begin to talk "for no reason."[46] In a sense I can agree with this. To say that someone did something "for no reason" is ambiguous. It can mean, on the one hand, that there were no causes of their doing it; or it can mean, on the other hand, that there is no R such that

[46] They actually say this with respect to fragility and irascibility, not loquaciousness, but surely mean it to apply to the case of loquaciousness as well.

the person's reason for doing it was R. Acting for a reason is tied to acting intentionally: if you do something for a reason, you do it intentionally.[47] But unintentional acts, acts not done for reasons, are not thereby uncaused. If a doctor hits my patellar tendon with a hammer, I kick my knee unintentionally, and so in a sense do it for no reason, but clearly the kicking has a cause, namely the hammer impact. There is no problem for my view if a loquacious person frequently begins talking for no reason in the second sense: there is nothing that is his reason for beginning to talk. You ask him why he started talking and he says "I don't know why" or "Not for any particular reason." That's compatible with its being the case that his arrival at the party caused him to begin talking. But we should not let the plausibility of the claim that there could be a loquacious person who frequently begins talking without there being anything that is his reason for beginning to talk carry over to the distinct claim that there could be a loquacious person who frequently begins talking without those beginnings being caused.

In addition to trying to imagine a loquacious person who frequently manifested this disposition in the absence of any triggering, Manley and Wasserman try, in effect, to run the argument in the opposite direction (with a different disposition): they try to imagine something that spontaneously exhibits the manifestation behavior of a given disposition D, and argue that that thing thereby has D. They imagine a kind of glass such that things made of it frequently break spontaneously. Surely, they implicitly claim, this would make things made of that kind of glass fragile (at least to some degree). It is important here that "spontaneously" means "uncaused"; it is not enough if the cause is microscopic. If an extra few atoms smash into a glass and thereby cause it to break, and we watch the glass break, we might say, since of course we couldn't observe the cause, that the glass broke spontaneously. But in this case the glass's breaking was not uncaused, and in fact this kind of spontaneous breaking would count in favor of

[47] The converse is controversial; maybe, as Anscombe held (1963), you can act intentionally but for no particular reason.

the glass's being fragile. The hard case is a kind of glass that frequently breaks where the breakings are truly uncaused. It seems to me that the fact that that kind of glass breaks spontaneously does not bear one way or another on whether the glass is fragile.[48] It could be fragile, or not; that would depend on what happens when you hit it.

One might reply, "okay, forget about whether the glass, in breaking spontaneously, manifests fragility. Surely it manifests some disposition—we could call it the disposition to break. Since it manifests this disposition in the absence of a trigger, there are triggerless dispositions." But this is not an argument, just an assertion that in breaking the glass manifests some disposition, an assertion I deny.[49]

XII

Besides dispositions, philosophers often say that mental states, like beliefs and desire, are causes of our actions. On my view they cannot be, for mental states are states and so not events.

How strong is the case that our beliefs and desires cause us to act? (I'll stick with beliefs and desires—what I would say about other mental states is analogous.) I'm not much bothered by off-hand comments that imply or presuppose that beliefs or desires are causes. What would worry me is if a decent theory of the mind implied that they are causes.

Then shouldn't I be worried? Speaking at a very high level of abstraction, isn't functionalism the thesis that beliefs and desires are defined by their causal roles? And doesn't the causal role of a

[48] I mean "does not bear metaphysically"; if the glass is fragile, it is not fragile partly in virtue of the spontaneous breakings. There could certainly be circumstances in which the spontaneous breakings were good evidence that the glass was fragile.

[49] The same goes for cases of particle decay. Molnar (2003, 85) observes that muons decay spontaneously (at least on some interpretations of quantum mechanics) into an electron, neutrino, and antineutrino. But it does not follow from this observation that in decaying they manifest some disposition. (Nor does Molnar argue that it does: his claim is that muons have the power to decay. I accept that.)

belief comprise facts about what causes that belief, and what that belief causes?

It is worth making explicit something that is implicit in these questions. It will not matter in what follows whether some functionalist theory of the mind is true. What matters is whether some functionalist theory correctly articulates the causal roles of mental states like beliefs and desire. For the question I want to answer is whether the causal role of, for example, a given belief, involves that belief actually causing anything, or being caused by anything. So long as some version of functionalism tells us the answer to this question, it does not matter whether the additional claim that functionalist theories make, that that belief is defined by its causal role, is true.

When you look at what functionalist theories say about the causal roles of mental states, you will find that those causal roles do not include being either causes or effects. Look at one canonical statement of functionalism, due to Ned Block in his well-known paper "Troubles With Functionalism":

> One characterization of functionalism that is probably vague enough to be accepted by most functionalists is: each type of mental state is a state consisting of a disposition to act in certain ways *and to have certain mental states*, given certain sensory inputs and certain mental states.
>
> (1978, 262)

According to this characterization, mental states like beliefs and desires are dispositions.[50] Set aside, as I said we should, the question of whether this identification is true, and look at what the theory says is the causal role of a metal state. It says that the causal role of, say, a belief is the same as the causal role of a disposition. This is compatible

[50] A more popular view, endorsed by for example David Lewis (1993), is that mental states are the bases of these dispositions. For an argument that mental states should be identified with dispositions, not with the bases of dispositions, see (Schwitzgebel 2013).

with my theory. For, as I have argued, dispositions are not causes or effects; instead they are background conditions to causings of effects.

Block's characterization of functionalism is, as he says, vague. More formal statements of functionalism are also consistent with my view. Here, for example, is a statement of "machine functionalism" (see for example Block 1978, 266–7). We need some definitions: a machine table is a list of conditional statements of the form:

> if Z is in state Si and receives input Ij then it goes into state Sk and produces output Ok.

Something X is a realization of a machine table if X has states Zi such that the machine table is true of X if the symbols "Si" are interpreted to denote the states Zi. The functionalist thesis then is that something has mental states if it realizes some (appropriate) machine table; and for each possible mental state, for example the state of believing that snow is white, if something has mental states, its state Zi is the state of believing that snow is white iff... (what follows lists what relations to inputs, outputs, and other states is definitive of this belief).

What matters for us is what the conditionals in the machine table look like. The conditionals are not couched in causal terms at all, but it is natural to understand them as making the following causal claims: inputs cause the machine to leave one state and go into another; and inputs cause the machine to produce outputs; and the state it is in is a background condition to both of these episodes of causation. Note that in the first episode the input does not cause a state, it causes the machine to go into a state. Since "go into state S" is non-stative, it corresponds to an event that can be caused. So this is a causal role for beliefs that is consistent with my claim that beliefs are states and therefore not causes or effects.

XIII

Throughout this chapter I've taken background conditions to always be background conditions to causings. But causation is analogous in many ways to grounding, so maybe on a more expansive understanding there are also background conditions to instances of grounding.[51] If there are, it would be interesting to investigate how well my theory of background conditions to causing translates into a theory of background conditions to grounding.

I'm not going to do that. But there is one debate about grounding that seems relevant and is worth a comment. In his theory of social ontology Brian Epstein distinguishes between the facts that ground a social fact and the "anchors" that in some sense underwrite the grounding of a social fact (2015). For example, Epstein holds that a given bill's being issued by the Treasury grounds that bill's being money, and that this grounding fact is anchored by the fact that we all accept that bills issued by the US Treasury count as money.[52] In Epstein's hands anchors seem to behave a lot like background conditions. Anchors underwrite instances of grounding in something like the way that background conditions underwrite instances of causing. Let's compare them. The striking of a match caused the match to light because the background condition (the presence of oxygen) obtained; the background condition enabled the cause to produce the effect. Similarly, it seems, the bill's being issued by the Treasury grounds its being money because the anchoring fact obtains; the anchoring fact enables the former to ground the latter. Maybe this is all the case because anchors are background conditions to instances of grounding. Maybe Epstein's theory of anchoring can be read as a theory of (one kind of) background condition to grounding.

Jonathan Schaffer has criticized Epstein's theory; he thinks that anchoring is just grounding under another name (forthcoming). But the parallels between anchoring and background conditions suggests responses to those criticisms. One of Schaffer's arguments uses the premise that anchoring behaves like grounding. I think his premise is false. Once you've signed on to thinking that causes and background conditions are different things, things that have different characteristic behaviors (that is, metaphysical roles to play), anchors look a lot

[51] One place where these analogies are explored is (Schaffer 2016).

[52] Actually, Epstein does not endorse a "collective acceptance" theory of the anchoring of social facts (Searle 1995 does endorse such a theory). But it is harmless here to pretend such a theory is true when discussing examples.

more like background conditions than causes. As I said in the last paragraph, anchors underwrite instances of grounding in the way that background conditions underwrite instances of causation.

Schaffer also claims that anchors are to grounding as Dretske's structuring causes are to causation; since structuring causes are a kind of cause, we should think of anchoring as a kind of grounding. The premise here is false too. I've argued, against Dretske, that structuring causes and background conditions are different kinds of things. And if anchors are analogous to one of them they are analogous to background conditions, not to structuring causes.[53]

The analogy between anchors and background conditions is enticing, but not perfect. A background condition has to obtain when and where the cause is doing its causing.[54] But on Epstein's view at least anchoring facts need not be local to the instances of grounding they underwrite: if US citizens all came to accept that sea shells count as money, then even if X is a sea shell far away on a foreign shore, our accepting that proposition anchors the fact that X's being a sea shell grounds X's being money. I'm not sure what to make of this difference.

XIV

A lot has happened in this chapter, so let me summarize my main claims. I have proposed a theory of background conditions that says that something is a background condition to C's causing E iff it is a state that is a reason why C caused E. This theory invokes the

[53] Maybe by "structuring cause" Schaffer does not mean what Dretske explicitly defined this term to mean (an explicit defintion that, I argued above, prevents "structuring cause" from applying to background conditions); maybe he wants it to have a meaning on which it definitely applies to background conditions, even if it then does not mean what Dretske defined it to mean. This does not affect my point: now Schaffer is right that anchors are like structuring causes, but he is wrong to say that structuring causes are a kind of cause.

[54] This is not an explicit part of my theory of background conditions, but it does, I believe, follow from the theory.

state/event distinction, which I draw by saying that an event occurs iff something engaged in an act, and a state obtains iff something instantiates a property. It is a consequence of the theory that having a disposition, and having a basis for a disposition, are states and so cannot cause things. Instead, I have argued, having a disposition is a background condition to the trigger's causing the manifestation of that disposition. Finally, I used this idea in an account of the conditions under which something manifests a disposition, rather than just exhibiting the manifestation behavior of that disposition.

References

Anscombe, G. E. M. 1963. *Intention.* 2nd edition. Harvard University Press.

Armstrong, David 1975. "Beliefs and Desires as Causes of Action." *Philosophical Papers* 4: 1–7.

Baierlein, Ralph 1999. *Thermal Physics.* Cambridge University Press.

Bennett, Jonathan 1988. *Events and their Names.* Hackett.

Bigaj, Tomasz 2005. "Causes, Conditions and Counterfactuals." *Axiomathes* 15: 599–619.

Bird, Alexander 2010. "Causation and the Manifestation of Powers." In Anna Marmodoro (ed.), *The Metaphysics of Powers.* Routledge, 160–8.

Block, Ned 1978. "Troubles With Functionalism." In C. Wade Savage (ed.), *Minnesota Studies in the Philosophy of Science Volume 9: Perception and Cognition.* University of Minnesota Press, 261–325.

Comrie, Bernard 1976. *Aspect.* Cambridge University Press.

Davidson, Donald 2001a. "Agency." In Davidson 2001b, 43–62.

Davidson, Donald 2001b. *Essays on Actions and Events.* 2nd edition. Oxford University Press.

Davidson, Donald 2001c. "Freedom to Act." In Davidson 2001b, 63–82. Originally published in 1973.

Dowdy, David R. 1979. *Word Meaning and Montague Grammar.* Kluwer.

Dretske, Fred 1988. *Explaining Behavior.* MIT Press.

Epstein, Brian 2015. *The Ant Trap.* Oxford University Press.

Fara, Michael and Sungho Choi 2014. "Dispositions." In Edward N. Zalta (ed.), *The Stanford Encyclopedia of Philosophy* (Spring 2014 Edition), URL= <http://plato.stanford.edu/archives/spr2014/entries/dispositions/>.

Frankfurt, Harry G. 1988. "The Problem of Action." In *The Importance of What We Care About.* Cambridge University Press, 69–80.

Hart, H. L. A. and Tony Honoré 1985. *Causation In The Law.* 2nd edition. Oxford University Press.

Hitchcock, Christopher and James Woodward 2003. "Explanatory Generalizations, Part II: Plumbing Explanatory Depth." *Nous* 37: 181–99.

Huddleston, Rodney 2002. "The Verb." In Rodney Huddleston and Geoffrey K. Pullum (eds.), *The Cambridge Grammar of the English Language.* Cambridge University Press, 71–212.

Hyman, John 2014. *Action, Knowledge, and Will.* Oxford University Press.

Johnston, Mark 1992. "How to Speak of the Colors." *Philosophical Studies* 68: 221–63.

Kenny, Anthony 2003. *Action, Emotion and Will.* 2nd edition. Routledge. 1st edition published 1963.

Kim, Jaegwon 1993. "Events as Property Exemplifications." In *Supervenience and Mind.* Cambridge University Press, 33–52.

Kment, Boris 2014. *Modality and Explanatory Reasoning.* Oxford University Press.

Lewis, David 1986a. "Causation." In *Philosophical Papers Volume II.* Oxford University Press, 159–71.

Lewis, David 1986b. "Events." In *Philosophical Papers Volume II.* Oxford University Press, 241–69.

Lewis, David 1994. "Reduction of Mind." In S. Guttenplan (ed.), *A Companion to the Philosophy of Mind.* Blackwell, 412–30.

Lewis, David 1999. "Finkish Dispositions." In *Papers in Metaphysics and Epistemology.* Cambridge University Press, 133–51.

Lombard, Lawrence Brian 1986. *Events.* Routledge & Kegan Paul.

Mackie, J. L. 1980. *The Cement of the Universe.* Oxford University Press.

Manley, David and Ryan Wasserman 2007. "A Gradable Approach to Dispositions." *The Philosophical Quarterly* 57: 68–75.

Manley, David and Ryan Wasserman 2008. "On Linking Dispositions and Conditionals." *Mind* 117: 59–84.

Mellor, D. H. 2004. "For Facts as Causes and Effects." In John Collins, Ned Hall, and L. A. Paul (eds.), *Causation and Counterfactuals.* Oxford University Press, 309–24.

Mill, John Stuart 1846. *A System of Logic.* Harper & Brothers.

Molnar, George 2003. *Powers.* Oxford University Press.

Mourelatos, Alexander 1978. "Events, Processes, and States." *Linguistics and Philosophy* 2: 415–34.

Nolan, Daniel 2015. "Noncausal Dispositions." *Nous* 49: 425–39.

Parsons, Terence 1990. *Events in the Semantics of English.* MIT Press.

Prior, Elizabeth 1985. *Dispositions.* Aberdeen University Press.

Prior, Elizabeth, Robert Pargetter, and Frank Jackson 1982. "Three Theses about Dispositions." *American Philosophical Quarterly* 19: 251–7.

Pullum, Geoffrey K. 1991. "English Nominal Gerund Phrases as Noun Phrases with Verb-Phrase Heads." *Linguistics* 29: 763–99.

Schaffer, Jonathan 2016. "Grounding in the Image of Causation." *Philosophical Studies* 173: 49–100.

Schaffer, Jonathan forthcoming. "Anchoring as Grounding: On Epstein's *The Ant Trap.*" *Philosophy and Phenomenological Research*, forthcoming.

Schnieder, Benjamin 2017. "The Naïve Conception of Properties." *Philosophical Issues* 27: 322–42.

Schwitzgebel, Eric 2013. "A Dispositional Approach to Attitudes: Thinking Outside of the Belief Box." In Nikolaj Nottelmann (ed.), *New Essays on Belief.* Palgrave, 75–99.

Searle, John 1995. *The Construction of Social Reality.* The Free Press.

Setiya, Kieran 2007. *Reasons Without Rationalism.* Princeton University Press.

Setiya, Kieran 2009. "Reasons and Causes." *European Journal of Philosophy* 19: 129–57.

Setiya, Kieran 2013. "Epistemic Agency: Some Doubts." *Philosophical Issues* 23: 179–98.

Setiya, Kieran 2016. *Practical Knowledge: Selected Essays.* Oxford University Press.

Skow, Bradford 2016. *Reasons Why.* Oxford University Press.

Smith, A. D. 1977. "Dispositional Properties." *Mind* 86: 439–45.

Steward, Helen 1997. *The Ontology of Mind: Events, Processes, and States.* Oxford University Press.

Strevens, Michael 2008. *Depth.* Harvard University Press.

Szabó, Zoltán 1994. "On the Progressive and the Perfective." *Nous* 38: 29–59.

Thomson, Judith Jarvis 1977. *Acts and Other Events.* Cornell University Press.

Vendler, Zeno 1967. *Linguistics in Philosophy.* Cornell University Press.

Vetter, Barbara 2015. *Potentiality.* Oxford University Press.

Wedgwood, Ralph 2007. *The Nature of Normativity.* Oxford University Press.

Weslake, Brad 2013. "Proportionality, Contrast and Explanation." *Australasian Journal of Philosophy* 91: 785–97.

Williamson, Timothy 2000. *Knowledge and its Limits.* Oxford University Press.

Wolterstorff, Nicholas 1970. *On Universals.* University of Chicago Press.

Yablo, Stephen 2008. "Mental Causation." In *Thoughts.* Oxford University Press, 222–48.

3

Dispositions: Intrinsicness and Agency

I

Are dispositions intrinsic (holding the laws fixed)? Certainly some of them are. Here is a fragile wine glass. Take it to your neighbor's house; dunk it in the bathtub; send it to the moon. It stays fragile, and will stay fragile through any change in its circumstances. To cause it to stop being fragile, you have to change the glass itself. But that's just one case, and plenty of philosophers doubt that every disposition is intrinsic.

The largest compendium of allegedly extrinsic dispositions is due to Jennifer McKitrick (2003).[1] Now we can go back and forth over the merits of each of her examples. But absent some more theoretical perspective on the debate, the back and forth is bound to be inconclusive. Maybe these examples fail to convince; but without some proposed reason why dispositions should be intrinsic, we cannot be very confident that better examples will not eventually appear. On the other side, those who reject the Intrinsicness Thesis are unlikely to convince their opponents simply by presenting examples.

[1] Among those who accept McKitrick's conclusion are Handfield (2008), Choi (2005), and Manley and Hawthorne (2005). The last authors' rejection of the Intrinsicness Thesis is, however, a bit hedged. The hedging is common: Manley and Wasserman, for another example, say that McKitrick's criticism of the Intrinsicness Thesis is "deserved," but they do not go so far as to explicitly reject the thesis (2008, 61). Barbara Vetter doesn't hedge, though: she says that McKitrick has "shown" that there are extrinsic dispositions (2015, 14n13).

It's not that you can never disprove a generalization by producing counterexamples; it happens all the time, the first observation of a black swan being just one instance. But alleged examples of extrinsic dispositions are much easier to "explain away" than are the observations that Willem de Vlamingh's expedition made on the Swan River in 1697. Not many people will think it reasonable to maintain that the animals Vlamingh saw (conspecifics of which we can also go see for ourselves) were not really black, or not really swans. It is less obviously unreasonable to say, of an apparent counterexample to the Intrinsicness Thesis, that the property that is supposed to be an extrinsic disposition is not a disposition at all, or does not even exist in the first place.

In this chapter I aim to provide a theoretical perspective on the Intrinsicness Thesis from which we can see why so many of McKitrick's examples are unconvincing, but also why her main thesis is correct: the Intrinsicness Thesis is false. That perspective comes from the necessary truth that a disposition must be a disposition to act.

II

The Intrisicness Thesis says that dispositions must be intrinsic (holding the laws fixed).[2] So whether something has a given disposition does not depend on its environment, its surroundings, on what things outside of it are like; it is only a matter of what that thing is like in itself (and what the laws of nature are).

Now if one is going to argue that a certain example shows that there exist extrinsic dispositions, there are three things that one must establish about the example. First, one must establish that the property

[2] Among those who accept this thesis are Johnston (1992), Lewis (1997), and Molnar (1999). The qualification about laws is there because many philosophers think that the thesis is obviously false without it: they think that anything could lose any of its dispositions, without changing intrinsically, if the laws of nature were different enough. Some views about laws give the thesis a chance of being true without the qualification; for example, dispositional essentialism (Bird 2007 defends this view).

the example concerns really does exist; second, that this property is a disposition; and third, that it is an extrinsic property. One of the first two tasks is going to be easier than the other, and which it is will depend on how you try to refer to the alleged disposition.

One way to try to refer to a disposition (or "semantically express" a disposition, if you reserve reference for nouns) is to use a dispositional adjective like "fragile" or "irascible." If you try to refer to a disposition by using an adjective like that, it will be easy to establish that the property of being ϕible/able exists. If you describe a scenario in which the adjective applies to something, that's proof that the property exists.[3] But you will have to do more work to show that the property is a disposition. Not every adjective of this form refers to a disposition.

Another way to try to refer to a disposition is to use an "explicit dispositional ascription" or "overtly dispositional locution."[4] That is a description of the form "the disposition to M when/in C," or a predicate of the form "is disposed to M when/in C," for example "the disposition to break when struck" and "is disposed to break when struck." If you try to refer to a disposition by using an explicit dispositional ascription it will be easy to establish that if the ascription denotes a property, it denotes a disposition. But you will have to do more work to show that the predicate does denote a property. Some predicates of this form don't denote to anything at all. Both of these obstacles to referring to dispositions stem from my thesis that dispositions must be dispositions to act, which I will defend in section IV. It will be useful to bear these obstacles in mind as we look at examples.

III

What are the allegedly extrinsic dispositions? In this section I will look at the battery of examples that McKitrick presents, and run

[3] I am waiving Russell-paradox style obstacles.

[4] These terms come from (Manley and Wasserman 2008, 60) and (McKitrick 2003, 157) respectively.

through some ideas about how the Intrinsicness Thesis may be defended against them. This discussion is meant to be suggestive but not decisive.

McKitrick discusses six examples that she claims are examples of extrinsic dispositions. The examples divide naturally into three groups:

Group 1: The disposition (of, say, a key) to unlock this door;[5] the disposition to dissolve the contents of my pocket (all I have in my pocket are some sugar cubes).

Group 2: Weight. McKitrick, following Yablo (2010, 51), identifies the property of weighing N pounds with "the disposition to depress a properly constructed scale so as to elicit a reading of N pounds in the local gravitational field."

Group 3: Vulnerability; visibility; recognizability.

Why think these are extrinsic? I'll take one representative from each group.

First, if there is such a disposition as the disposition to unlock this door, it is extrinsic; a key could lose this disposition without changing intrinsically, if we fuss with the lock on the door. Second, if weighing N pounds is a disposition then it is an extrinsic disposition; I weigh more on Earth than on the Moon. Third, visibility. McKitrick distinguishes two different uses of the word "visible." The first is the use it has in sentences like "Tables are visible objects," or "The Empire State Building is visible, not like helium gas, which is invisible." She does not claim that "visible" denotes an extrinsic disposition, on this use. Her claim is about the use it has in sentences like "Saturn is going to be visible tonight for the first time this year" (162). There are times in the year when Saturn is not visible, in the second sense, even though it is always visible in the first sense. Let us from now on interpret "visible" and related words, like "visibility," according to the

[5] A similar example was discussed by Shoemaker (1984, 221); I will say something about the differences later.

second sense. McKitrick identifies visibility, in the second sense, with the disposition to be seen when someone looks toward it. Now if there is such a disposition as this one, and if this disposition is identical to the property of being visible, then this disposition is extrinsic; a far off mountain may be visible on a clear day, and not on a foggy one, without undergoing any intrinsic change.

Doubts can be raised about each example. Start with the key. The key certainly has the power to unlock the door. But although they are often lumped together, powers and dispositions are different things. Having the power to unlock the door is not sufficient for being disposed to unlock the door.[6] If I have the key then I have the power to unlock the door. But if there is something dangerous or scary behind the door, I might really want the door to stay locked. Then although I have the power to unlock the door, I am not disposed to unlock the door.

These worries about the key might seem beside the point. Maybe the key has the power to unlock the door without being disposed to unlock the door. So what? The real question is whether there is such a disposition as the disposition to unlock this door. The fact that if there is such a disposition, the key does not have it, does not show that the disposition does not exist. Maybe there are better candidate bearers of the disposition? I will serve. Here is the key, in my hand; there is the door. I have formulated the policy of unlocking this door whenever I see someone is on the porch. Am I not thereby disposed to unlock this door (with the key) whenever I see someone is on the porch? Suppose I am; then there is such a disposition as the disposition to unlock this door (when I see someone on the porch). And surely this disposition is extrinsic. I would lose it, without any change in my policies, if the lock on the door were changed.

There is still room for doubt. Am I really disposed to unlock the door? Maybe I'm just disposed to try to unlock the door. This disposition doesn't go away when the lock on the door is changed. If the

[6] While McKitrick gets the key example from Shoemaker, Shoemaker attributes to the key the power to unlock the door, not the disposition to unlock the door.

lock hasn't yet been changed, then when I manifest my disposition to try, I succeed in unlocking the door; but, the thought goes, that's not enough for me to be disposed to unlock the door.

I grouped the disposition to unlock the door with the disposition to dissolve the contents of my pocket. One might wonder how representative the disposition to unlock the door is of its group. Do the doubts I have raised about whether it is extrinsic correspond to any doubts about the disposition to dissolve the contents of my pocket? They do. The corresponding doubt is the thought that a bucket of water is not really disposed to dissolve the contents of my pocket (when "the contents of my pocket" is given narrow scope); instead it just has the intrinsic disposition to dissolve these sugar cubes (the ones in my pocket). (In other words: the dispositional ascription is only true when "the contents of my pocket" is given wide scope.) In both cases the thought is that the thing in question (me, the water) only seems to have an extrinsic disposition, when really it just has an intrinsic disposition that is easily confused with it.

What about weight? Is weight really a disposition? Again, McKitrick identifies weighing N pounds with the disposition to "depress a properly constructed scale so as to elicit a reading of N pounds in the local gravitational field." Now each disposition has "manifestation conditions" and "stimulus conditions," descriptions of which go in for "M" and "C" respectively in "is disposed to M when/in C." The stimulus conditions for this disposition are not completely explicit; let's make them explicit. The thesis is that weighing N pounds is (identical to) being disposed to depress a properly constructed scale so as to elicit a reading of N pounds, when placed on such a scale in the local gravitational field.[7] Now, I think this identification is false. I think that the property of weighing N pounds is the property of being such that there is just one massive body much much larger than

[7] We are supposed to interpret "local" in "local gravitational field" the way we interpret "local" in "Everyone went to a local bar," when we meant that each person went to a bar near that person. "X was placed on a scale in the local gravitational field" is true iff X was placed on a scale near X, whether or not the scale is near the person who uttered the sentence. More on this later.

you nearby, and that body is exerting a gravitational force of N pounds on you.[8] This property is extrinsic: you can lose it without changing intrinsically, if you move from the neighborhood of one much larger massive body to another (one that has a different mass). But this property isn't a disposition. Even if you doubt this particular theory of weight, you should, I think, doubt that weighing N pounds is identical to being disposed to elicit a reading of N pounds on a scale (etc.). For surely things that weigh N pounds are disposed to elicit readings of N pounds on scales because they weigh N pounds. But their weighing N pounds cannot explain why they are disposed to elicit N-pound readings if their weighing N pounds just is their being so disposed; nothing can explain itself.

Again, though, debates about what weight is are just preliminary skirmishes. We have the overtly dispositional locution "is disposed to depress a properly constructed scale so as to elicit a reading of N pounds when placed on such a scale in the local gravitational field." Forget about whether this denotes the property of weighing N pounds; it certainly denotes a disposition. Isn't this disposition extrinsic?

This question presupposes that there is one single property that this predicate denotes in every context. This is controversial. The predicate contains "local," a context-sensitive word, and the context-sensitivity of this word presumably makes the dispositional predicate containing it context-sensitive as well. McKitrick tries for an interpretation on which it is not context-sensitive: she insists that "local" always be interpreted to mean "near the thing to which the predicate is applied" rather than "near the speaker"; her claim is that the predicate so-interpreted denotes the same property in every context (if it denotes any property at all). Maybe this is right, but it would be better to avoid the controversy entirely. We can do this by focusing on a similar example that raises no questions about context-sensitivity, namely the disposition to elicit an N-pound reading when placed on a scale on the surface of the Earth. The predicate "...is disposed to elicit an

[8] This characterization could certainly be improved, but will do for my purposes.

N-pound reading when placed on a scale on the surface of the Earth" is not context-sensitive (in any relevant respect); and if it denotes anything, it denotes an extrinsic disposition. You can make something gain or lose the disposition just by changing the mass of the Earth.

Does this predicate denote anything? If it does, then a rock can satisfy the predicate. Place a rock weighing N pounds on a properly constructed scale. Assuming everything goes right, the scale will then read N pounds. If the rock is manifesting the disposition to depress scales so as to elicit N-pound readings, then the rock is depressing the scale and eliciting a reading from the scale. One might doubt that these are things that the rock is doing. Can rocks do things in the first place? If not, then our rock is not doing these things, and if it is not doing these things, then it is not manifesting a disposition to do these things.

Let me leave weight for now, and turn to the third class of examples. Having spent so long on the first two classes of examples, I will be brief here. As with weight, there is room to doubt that visibility is a disposition. I think that the property of being visible is the property of being such that people are able to see you. But this property doesn't look much like a disposition.

As before, one could shift from arguing that visibility, which is obviously extrinsic, is a disposition, to arguing that the disposition to be seen when looked at, which is obviously an extrinsic disposition if it exists, really does exist. But there is something funny about the predicate "...is disposed to be seen when looked at." Is it really well-formed? If not, it doesn't denote anything, so doesn't denote a disposition.

None of this is satisfying. We need a broader theoretical perspective on the debate. I have one: it comes from recognizing a necessary connection between dispositionality and agency.

IV

The idea is this: it is a necessary truth that every disposition is a disposition to act.

By an act I just mean a thing something/someone does. Running is an act; it is something I do fairly often. Being tired, on the other hand, is not an act. It is not something I ever do, it is not something that anyone could do—not because being tired is some kind of "impossible act," like squaring a circle, but because it is not an act at all. If it were an act, it would at least make sense to say "one thing he did was be tired"; but this does not make sense. (Tired is something I often am; but I can be tired without *being tired* being something that I ever do.)

In this argument I've used a test for when a verb or verb phrase denotes a type of act: if "one thing he/she/it did was V" makes sense, is a grammatically well-formed sentence, then the verb phrase going in for "V" denotes a type of act. If acts are things that may be done, then this is obviously a good test.

I put "it" in as an option for this test because things that have no sex can still perform acts. When a plant closes its petals, the plant does something, namely close its petals.

Agency goes with action: an agent is something or someone that is able to perform acts. Since plants can do things, plants are agents. Some might balk at this; but the source of resistance is probably the tradition in the theory of action of using "act" and "agent" in more demanding senses than mine. Often "act" is used to mean "intentional act" or "act done for reasons," and "agent" is used to mean "thing that can act" in this new sense of "act."[9] Plants can't act for reasons. Although these are perfectly legitimate uses of "act" and "agent" their existence does not show mine to be illegitimate. In fact their legitimacy seems to prove the legitimacy of mine: if the notion of an act that can be done for reasons makes sense, then surely so does the notion of an act that is not done for reasons. Similarly, if the notion of a thing that can act for reasons makes sense, then surely so does

[9] One example of this is Dretske's (1988). He opens by contrasting "things we do" with "things that happen to us," and uses "behavior" as a general term for things we do, reserving "action" for only a certain subset of the things we do. I mean by "action" what Dretske means by "behavior."

the notion of a thing that can act, but may not have the capacity to act for reasons.

I've said that a verb phrase denotes an act if it makes sense to use it to complete "One thing he did was..." Now linguists divide verbs into "aspect classes," the most basic division being that between stative verbs and non-stative verbs. And one way to test whether a verb is non-stative is the same as my test for when it denotes an act: check whether it makes sense in "one thing he did was..." So my claim becomes the claim that non-stative verbs denote acts, and stative verbs do not.[10]

Whether a verb phrase passes the "one thing he did was..." test depends on the form of the verb phrase used. "Stabbing Smith" is the progressive form of "Stab Smith," and "one thing Jones did was stabbing Smith" is not grammatical; this does not show that "stab Smith" is stative. The test is to be applied to the verb phrase in its so-called "plain" form.

This dependence on the form shows that it is over-simplified to say that satisfying a non-stative verb phrase is sufficient for acting. Doing something is sufficient for acting, but satisfying a non-stative verb phrase is not sufficient for doing something. Clauses in the passive voice are a counterexample: If "Smith was stabbed by Jones" is true, then Smith satisfies "...was stabbed by Jones," which contains "stab," which is non-stative, but it is false that Smith did something. What is sufficient for doing something is satisfying (being denoted by the subject of) an active non-stative verb phrase.

Why do I hold that dispositions must be dispositions to act? Because it explains why certain phrases with the form of explicitly dispositional predicates are not grammatical. Predicates of the form "is disposed to M when C" are not grammatical unless (i) the verb phrase that goes in for "M" is non-stative, and so denotes an act-type, and (ii) that verb

[10] In taking all non-stative verbs to denote acts I follow Setiya (2013). See (Dowdy 1979) for a battery of tests, including this one and the "may appear in the progressive" test that I use later, for distinguishing stative from non-stative verbs.

phrase is in the active voice. "X is disposed to run when scared" and "X is disposed to open its petals in the morning" are grammatical, and "run" and "open its petals" are active forms of non-stative verbs. But "X is disposed to be red in the morning" and "X is disposed to be opened at noon" are not well-formed, and "be red" is stative and "be opened" is passive. (If you hesitate about "X is disposed to be red in the morning," it may help to distinguish it from "X is disposed to *turn red* in the morning." This is fine, and fits with the rest of the examples: "turn red" is a non-stative verb phrase.)

I've given only a couple of examples, but it is easy to run through many more in your head and see that they all fit the pattern. "Is disposed to M when C" is grammatical only when the verb phrase that goes in for M is non-stative and active. But satisfying a non-stative verb phrase in the active voice (that is, being denoted by the subject of a true clause with a non-stative active verb phrase) is sufficient for acting. My thesis explains this data. If dispositions didn't have to be dispositions to act, we should be able to form predicates in English for some of the non-act dispositions, by putting a stative verb in for "M." Since we cannot, dispositions must be dispositions to act.

Further confirmation that only non-stative verbs can go in for "M" comes from the fact that all instances of the following schema are true:

> If X is disposed to M in C, it has the power to M in C.

Nothing can be disposed to do what it lacks the power to do. But it is obvious that "power to M" is only grammatical if what goes in for "M" is non-stative and active. The power to M must be the power to act.

Here is a similar bit of evidence. Michael Fara claimed that the dispositional ascription "X is disposed to M in C" entails the habitual "X Ms in C" (2005). For example: "This barrel is disposed to roll when pushed" entails "This barrel rolls when pushed." He wanted to use this idea to build an analysis of dispositions in terms of habituals. The details of his analysis aren't important here. What's important is what follows from the conjunction of his claim and the claim (1) that habituals must have active non-stative verb phrases in for "M."

To establish (1) we need to know what habituals are. Unfortunately it is not easy to give a definition. One rough-and-ready definition says that a present-tensed non-progressive sentence is an habitual iff it reports what something frequently, or normally, or usually does, rather than what it is currently doing.[11] Using this definition we can establish (1) by checking cases. Here are two: "this barrel rolls when pushed" is an habitual. It can be true even if the barrel is not rolling. And "rolls when pushed" is active and non-stative; "one thing the barrel did was roll" is grammatical. On the other hand, "This barrel is round" is not an habitual, and "is round" is stative.

If Fara's claim and claim (1) are true then dispositions must be dispositions to act.[12] Let "X is disposed to M in C" be any explicit dispositional ascription. By Fara's claim it entails the habitual "X Ms in C." If one sentence entails another then both are grammatically

[11] This is only a definition of "present-tensed habitual"; "This barrel used to roll when pushed" is a past-tensed habitual. Dowdy discusses the viability of this definition in his (1979, 56). For an argument that an habitual "X Ms when C" can be true even when X does not normally M when C, see (Fara 2005, 65). For a definition that does not invoke what normally or frequently or usually happens see (Comrie 1976, 27).

Technically, we must speak of habitual interpretations of sentences, rather than habitual sentences. Without special stage-setting the most natural interpretation of "Jones builds houses" is as an habitual; it can be true (on that interpretation) at times when Jones is sleeping. But in some contexts, for example a story told in the present tense, the most salient interpretation of "Jones builds houses" is non-habitual, reporting a particular episode of building: "It is Tuesday. Jones wakes up, puts on his clothes, and builds houses until lunchtime. He does this even though he is not a house-builder at all; it is a one-off thing he is doing for charity." With this complication taken into account, my thesis is that a present-tensed (non-progressive) sentence has an habitual interpretation only if the main verb is non-stative.

[12] Some think that there are counterexamples to Fara's claim involving masked dispositions. A fragile glass is packed in bubble-wrap. Jones drops it over and over again, trying and failing to break it, finally saying in frustration, "This glass does not break when dropped!" If he speaks the truth, and if the glass remains fragile while packed in bubble-wrap (a widely accepted claim), this is a counterexample (Manley and Wasserman 2008, 73). Fara replies that "This glass is disposed to break when dropped" is context-sensitive, and that in Jones' context it is false (2005, 76).

well-formed. So the habitual "X Ms in C" is grammatical. And if it is grammatical then by (1) it has an active non-stative verb phrase in for "M." It follows then that "X is disposed to M in C" also has an active non-stative verb phrase in for "M." If every disposition may be referred to by such an ascription, then dispositions must be dispositions to act.

One might object to this conclusion, claiming that there are examples of dispositions that are not dispositions to act. What about, for example, the disposition to be talkative at parties? There doesn't seem to be anything ill-formed about "is disposed to be talkative." But "is talkative," like any predicate of the form "*be adjective*," is a stative verb phrase.

Actually it is false that "*be adjective*" is always stative. This is most easily seen using another test for non-stativity, the "may appear in the progressive" test. For the most part, non-stative verb phrases may appear in the progressive aspect, while stative verb phrases may not. And "John is being talkative" is perfectly grammatical. Since "is talkative" has a non-stative use, the fact that "is disposed to be talkative" is grammatical is compatible with my thesis (provided that "be talkative" is used non-statively in it). The same goes for many other instances of "*be adjective*," for example "is lazy/violent/kind."

To be clear, I do not deny that "is talkative" is sometimes stative. It has both a stative and a non-stative use. I can say in one breath "Jones is talkative, but he's not being talkative right now." The first occurrence is stative, meaning roughly that it doesn't take much to get Jones to talk, and the second occurrence is non-stative, meaning roughly that Jones is not acting talkative right now.[13]

Another possible objection to my thesis comes from basic physical properties. Some philosophers hold that all basic physical properties are dispositional. Mass and electric charge, for example, are, on this

[13] Parsons calls "be" when it appears in non-stative constructions like "Jones is being talkative" "the 'be' of action," since "Jones is being talkative" means roughly the same as "Jones is acting talkative" (Parsons 1990, 35; see also Comrie 1976, 36).

view, dispositional properties. But one might claim that they cannot be dispositional, if dispositions must be dispositions to act.

This objection doesn't stand up to scrutiny. A natural view is that if *having mass N kg* is a dispositional property it is identical to the following disposition (assuming Newtonian gravitational theory is true):

> The disposition to (i) accelerate at $F/N\ m/s^2$ when experiencing a force of F Newtons, and (ii) pull on bodies with mass M kg with a force of $G\frac{MN}{r^2}$ Newtons when at a distance of r meters from them.

Since "accelerate" and "pull" are non-stative, accelerating and pulling are things that can be done, so the existence of this disposition is consistent with my thesis.

The objection might linger: even if accelerating and pulling are things that can be done, they are not things that protons, neutrons, electrons, and so on actually do. Protons etc. are not agents; while things happen to them, while they are "involved" in the occurrence of various events, and while these happenings and events happen and occur in accordance with the laws of physics, protons etc. are entirely passive.

This is an idea that has had its adherents. It seems to be part of some versions of a "Humean" way of thinking about reality, and is also allied to the doctrine of occasionalism, which says that material bodies never act on each other, that instead the only actor in the physical world is God.[14] But a proponent of the idea that mass, charge, and so on are dispositional properties should just reject this view. Theirs is a view on which fundamental particles are active.

[14] It is probably no coincidence that both Humeanism and occasionalism show up here, since Hume's theory of causation has been called "occasionalism without God."

I said that the thesis that protons etc. are not agents is "allied" to occasionalism, rather than that it is part of occasionalism, because occasionalism is consistent with the claim that material bodies act; something can act without acting on something else. When a proton pulls on another protonit

The dispositionalists would be at a disadvantage if their view required some drastic change in the way non-philosophers think and talk about fundamental particles. I imagine someone might claim that it does: "Isn't the idea that fundamental particles are agents a drastic departure from what those not exposed to philosophy think?" Actually, I think it's not. Maybe if you asked "Do you think that protons are agents?" you would often get no as an answer, but this is too loaded a way to ask the question. "Agent" is too easily interpreted as "agent capable of acting intentionally," which certainly protons cannot do. The intended question, though, is whether protons ever do anything; and instead of focusing on such a general question, it is better to focus on a particular thing protons might do, and see if it is startling to admit that they do that thing. Here's a particular thing protons might do: accelerate. Accelerating is a thing that can be done; it passes the test. And if we ask those not contaminated by philosophy whether protons ever accelerate, whether accelerating is something a proton can do, I think the most common answer will be yes.

In fact, if we focus just on how well a theory fits with ordinary ways of thinking, it is those who deny that protons are agents, those who deny that protons do anything, who are at a disadvantage. They must hold that no proton has ever accelerated, since if a proton has accelerated, then one thing it has done is accelerate. This view, then, calls for linguistic reform, or at least linguistic innovation. A completely accurate description of the motion of a proton cannot use the verb "accelerate."

acts on that other proton, but when it accelerates it acts without acting on something else.

A contemporary Humean will probably not want to deny that protons act, if accelerating is sufficient for acting. A fall-back Humean view is the view that the most fundamental description of reality will not say that anything acts; "Proton X is accelerating" gives way at a deeper level of description to "Proton X is here at t1, there at t2, . . . ," a statement that uses only stative verbs. It is a good question whether the non-stative can always be reduced to the stative, but not one I will take up here.

V

The claim that dispositions must be dispositions to act helps cut through the debate over whether dispositions must be intrinsic.[15] It shows that, and why, many of McKitrick's examples fail, and also helps shore up the case that some of them really are examples of extrinsic dispositions.

Start with visibility. McKitrick, again, identifies visibility with disposition to be seen when someone looks toward it. But this cannot be right. While there is such a property as visibility, there is no such property as the property of being disposed to be seen when someone looks toward you. The verb phrase "is disposed to be seen" is ungrammatical both because "see" is stative and because it occurs in this phrase in the passive.[16] Dispositions must be dispositions to act, but in being seen one is not acting. The right thing to say about visibility is what I said about it in section III: to say that something is visible is not to ascribe to it any dispositions, but instead to ascribe to certain others (normal observers) a certain ability: the ability to see it.

The same goes for the rest of McKitrick's Group 3 examples. Vulnerability is not, as she claims, the disposition to be harmed if attacked, for there is no such disposition: "be harmed" is passive. In being harmed one is not acting. To say that a city is vulnerable is not to ascribe to it any dispositions, but instead is (just) to ascribe to certain other things—enemies of the city's occupants, say—a certain ability: the ability to harm it. And recognizability is not, as she claims, the disposition to be recognized: "be recognized" is passive. In being

[15] I'll also note in passing that my claim scuppers Barbara Vetter's proposal to ground possibility in dispositionality (2015). Her project requires the intelligibility of dispositional locutions of the form "X is disposed for it to be the case that P"; roughly, her view is that it is possible that P iff some X is disposed for it to be the case that P. But if my claim is right, no instance of this form could be true.

[16] Of course "see" is non-stative when used to mean "date," but that is not what it means here.

recognized one is not acting. To say that someone is recognizable is to ascribe to other people the ability to recognize it.

Despite the resemblance between the words "fragility," on the one hand, and "visibility" and "vulnerability" on the other, fragility does not belong with visibility and vulnerability. True, just as to say that something is visible is to say that others are able to see it, to say that something is fragile is to say that others are able to break it. But in the case of attributions of fragility, more is being done than attributing to others a certain ability. That's because, in the case of fragility, when others exercise that ability, the ability to break the glass, they cause the glass to do something. When someone exercises their ability to break the glass, they cause the glass to break, and when they cause the glass to break, the glass breaks: that's what the glass does. That's why attributions of fragility also attribute to the glass a disposition to break. None of this follows in the case of visibility. When others exercise their ability to see Saturn, they do not cause Saturn to do anything; certainly they do not cause it to "be seen."

I've said that fragility does not belong with visibility and vulnerability; but surely the common suffix "-ility" suggests that they do belong together? I learned the right answer to this question from John Maier's paper "Dispositions and Ergativity" (2015). Switch from the nouns ("visibility") to the corresponding predicates ("is visible"). Each of these predicates is semantically related to a transitive verb: "fragile" to "break," "visible" to "see," "vulnerable" to "harm." They are semantically related in this way: "If X is fragile, it can be broken," "If X is visible, it can be seen," and so on, are analytic. This similarity in the predicates is the way in which they belong together. But it does not follow that all the predicates are dispositional predicates. Maier argues, in fact, that a predicate "is X-able" is a dispositional predicate only if the verb semantically related to "X-able" is an ergative verb, where a verb is ergative iff (i) it has both transitive and intransitive forms, and (ii) "X $V_{\text{transitive}}$ Y" entails "Y $V_{\text{intransitive}}$." This sorts the cases as I have sorted them: "break," the verb related to "fragile," is ergative: if X broke Y, then Y broke, so "fragile" satisfies the necessary condition on being dispositional. "See," on the other hand, does not meet this condition: it is not ergative, for it is false that if Jones saw The White House, The White House saw.

It is worth noting that Maier's necessary condition on "is X-able" being dispositional entails mine, provided that ergative verbs are non-stative.

VI

The disposition to elicit an N-pound reading when placed on a scale on the surface of the Earth is an extrinsic disposition if it exists; does it exist?

The kind of argument I've been giving that certain dispositions do not exist does not work here. *Depressing a scale* and *eliciting a reading* are things that can be done. The disposition to depress a scale so as to elicit a certain reading is a disposition to act. And this disposition is extrinsic.

Now when I discussed this example earlier I expressed some reservations about whether a rock could be disposed to elicit readings on scales, doubts that stemmed from doubts about whether *depressing a scale* and *eliciting a reading* were things a rock could do. But what I've said since then about agency should dispel these doubts. We've already agreed that mindless material bodies can exert forces. But then there is something the rock can do such that, by doing it, the rock depresses the scale. It can depress the scale by exerting a force on the scale. N-pound rocks, then, really are disposed to elicit N-pound readings from scales when placed on them on the surface of the Earth; so they have an extrinsic disposition; so there are extrinsic dispositions. The Intrinsicness Thesis is false.

But this is not the end of the story.

VII

When one learns that the Intrinsicness Thesis is false one naturally wonders whether there is some natural fallback thesis. Maybe dispositions of a certain kind are always intrinsic. Or maybe every extrinsic disposition is in some way derived from an intrinsic one.[17]

One idea comes from Barbara Vetter (2015, §4.4).[18] Start with the disposition to elicit N-pound readings from this (particular) scale when placed on it on the surface of the Earth. This disposition is extrinsic, because a rock can lose it without changing intrinsically, if for example the mass of the Earth is changed. But while a change in the mass of the Earth is not an intrinsic change in the rock, it is an intrinsic change in the mereological sum of the Earth and the rock (let's assume that this sum exists). This suggests investigating the hypothesis that the rock's extrinsic disposition to elicit N-pound readings is somehow derived from an intrinsic disposition of the "Earth + rock," and also investigating the general hypothesis that every extrinsic disposition of a thing is derived from an intrinsic disposition of a larger thing (a thing that has the first thing as a part).

Let's investigate the general hypothesis by investigating the specific one. There are two questions: first, is there a disposition of the Earth + rock from which the rock's disposition to elicit N-pound readings is derived? Second, if so, is that disposition intrinsic? As to the first question, one might suggest that the Earth + rock is disposed to elicit N-pound readings from the given scale when its rock-part is placed on that scale near the surface of its Earth-part; maybe the rock's disposition is derived from this disposition. Does the Earth + rock really have this disposition? If it does, then the Earth + rock in manifesting it elicits an N-pound reading from a scale. But one might think that only the rock elicits this reading, not the Earth + rock. I will set this issue aside, and grant that the sum does have this disposition.

[17] McKitrick imagines an opponent suggesting that "[i]n the purported examples of extrinsic dispositions, some underlying intrinsic dispositions are really doing the work" (167), and later also imagines an opponent suggesting that every extrinsic disposition is "derived from" an intrinsic disposition (168). She does not object to these ideas, or even evaluate them in any detail; she is content to point out that, even if true, they do not show the Intrinsicness Thesis to be true.

[18] Her focus is on potentialities, not (just) dispositions, so the way I will present the idea isn't exactly her way.

I will also set aside, for now, the question of how the disposition of the rock is supposed to be derived from the disposition of the sum, and take up the question of whether this disposition (of the Earth + rock) is intrinsic. It is not, for a variety of reasons. If Saturn were to move much closer to the Earth, the Earth + rock would lose the disposition, because Saturn's gravitational pull on the rock would significantly alter the net force it experiences. And there's nothing special about Saturn here: if any very massive body, or collection of bodies that together are very massive, were to get close, the Earth + rock would lose the disposition. If there is something with the rock as a part that has an intrinsic disposition to elicit N-pound readings when its rock part is put on the scale, it is probably the mereological sum of all material bodies.

Back to the question of how the disposition of the rock is supposed to be derived from the disposition of the sum. One natural suggestion is that derivation is explanation: the rock has its disposition because the rock + Earth has its disposition.

I think this explanatory claim is false. In general, explanatory dependencies run from smaller things to larger things. So if there is an explanatory connection between a disposition of a rock and a disposition of the sum of that rock and the Earth, then the sum has its disposition in part because the rock has its, not the other way around. Since this explanatory claim is part of the hypothesis, the hypothesis itself is false.

VIII

I have another fallback version of the Intrinsicness Thesis to propose. To find a fallback version we need to find a sense in which intrinsic dispositions are more basic than extrinsic ones. There are many senses in which one thing might be more basic than another; which sense(s) can we plug in here to get a truth?

The word "basic" itself is a clue. The other clue is the thesis I started with: that dispositions must be dispositions to act. In the theory of action there is a distinction between basic and non-basic acts. Maybe that distinction can be used to generate a distinction between basic and non-basic dispositions. And then maybe the basic dispositions are all intrinsic.

Let's give it a try. First, the distinction between basic and non-basic acts. I do something as a non-basic act when I do it by doing something else, as when I hail a cab by raising my hand. I do something as a basic act when I do not do it by doing something else.[19] Usually discussions of basic and non-basic acts focus on examples of human acts, but once we accept that things like rocks can do things, we should accept that some of their acts are basic, and some non-basic. For example, eliciting an N-pound reading on a scale is a non-basic act of a rock's. It elicits an N-pound reading by exerting a force of N pounds on the scale.

When our rock manifests its disposition to elicit N-pound readings it does so as a non-basic act, because it elicits the reading by exerting an N-pound force. Here is a natural question: when the rock does the basic act, when it exerts an N-pound force on a scale, is it also manifesting a disposition? I think the answer is yes. Newton's third law says that when one body exerts a force on another, then the other exerts a force on the first, equal in magnitude and opposite in direction. It follows from this law that each material body is disposed to exert an N-pound force (in a given direction) on another body when that other body exerts an N-pound force on it (in the opposite direction). Now when you put the rock on the scale, the Earth exerts an N-pound gravitational force on the rock, and—since the rock stays on the scale—the scale also exerts an (upward) N-pound force on the rock. This triggers the rock to manifest its disposition to exert an N-pound (downward) force on the scale.

The interesting thing about all this is that the disposition to exert an N-pound force on something when it exerts an N-pound force on you is an intrinsic disposition. Holding the laws fixed, everything (in this

[19] The distinction between basic and non-basic acts was introduced by Danto (1965). There is more than one distinction that goes under this name; see Hornsby (1980, chapter 5).

and other possible worlds) has it, so in particular every duplicate of every thing has it. And that means that the rock manifests an extrinsic disposition, the disposition to elicit N-pound readings from scales on Earth, by manifesting an intrinsic disposition.

This suggests the following Revised Intrinsicness Thesis:

> Whenever something manifests an extrinsic disposition, it does so by manifesting an intrinsic disposition (or by manifesting a sequence of intrinsic dispositions).

I don't know of any counterexamples to this thesis.[20]

Could my disposition to unlock this door when I see someone on the porch be a counterexample to the Revised Thesis? I earlier suggested that this is not even a counterexample to the original Intrinsicness Thesis. The suggestion was that I do not have this disposition in the first place (in the hypothetical scenario), that I have instead the disposition to try to unlock the door. But when the distinction between basic and non-basic acts is near the front of our minds this suggestion is implausible. Surely I have both dispositions.

To try to do X is to do something in order to do X (McCann 1975). It is to do something that is, by your lights, a means to doing X, and to do it because it is a means. It follows that to be disposed to try to do X, there must be some acts Z such that you believe that doing any of Z is a means to doing X, and you are disposed to do-one-of-Z-in-

[20] The Revised Thesis does not say that whenever something has an extrinsic disposition, it also has an intrinsic one, and manifests the first by manifesting the second. This is a stronger claim, though the scenarios in which they come apart are complicated. The stronger claim would be falsified by a scenario in which (i) something X is disposed to M in C, (ii) this disposition D is X's only disposition, and (iii) D is extrinsic. But a possibility of this kind is not (yet) a counterexample to the Revised Thesis. A possibility of this kind is compatible with the Revised Thesis if the following is also true of it: when a triggering event for D happens, X acquires an intrinsic disposition, and it then manifests D by manifesting the intrinsic disposition. (In technical terms, the intrinsic disposition here is a reverse-fink. Finks and reverse-finks were so-named by C. B. Martin (1994).) The Revised Thesis as stated is a condition on manifesting an extrinsic disposition, not on having an extrinsic disposition.

order-to-do-X. For simplicity suppose there is just one Z. Then if in fact doing Z is a way to do X (in some concrete situation), and you are disposed to do Z in order to do X, then when you manifest this disposition in that situation, you succeed in doing X. Surely if all this is true then you are also disposed to do X.

In the scenario we're talking about I don't have false beliefs about how to unlock the door. As long as the lock has not been changed, when I try I succeed. So if I am disposed to try to unlock the door, it seems that I am also disposed to in fact unlock the door (when someone is on the porch).

But, again, when I manifest the disposition to unlock the door, I do it by manifesting the disposition to take some means to unlocking the door. This example is consistent with the Revised Thesis if the disposition to take the means is intrinsic. Is it? Well, I have to move my arm to unlock the door. So the means involves moving my body. But mustn't a disposition that I cannot manifest without moving my body be extrinsic? Consider a duplicate of me encased in rock up to his neck. He sees someone on the porch, but he cannot, and so does not, move. He doesn't manifest a disposition to try. Mustn't that be because he lacks the disposition to try? If so, that disposition is not intrinsic.

This argument is bad. My duplicate is disposed to try to unlock the door; this disposition is just masked. If a glass can remain fragile even when packed in styrofoam, I can retain the disposition to try to unlock the door when encased in rock. The example of the disposition to unlock the door does not refute the Revised Thesis.

IX

I've argued that dispositions must be dispositions to act, and that this thesis lets us reach a more definitive view about whether dispositions are intrinsic. Many proposed examples of extrinsic dispositions fail because the dispositions in question do not even exist:

if they did, they would not be dispositions to act. Some proposed examples of extrinsic dispositions, on the other hand, succeed: the disposition to elicit N-pound readings when placed on a scale on the Earth is one. But the structure of this example suggests a revised version of the Intrinsicness Thesis, a thesis that appears to be defensible.

References

Bird, Alexander 2007. *Nature's Metaphysics.* Oxford University Press.

Choi, Sungho 2005. "Do Categorical Ascriptions Entail Counterfactual Conditionals?" *Philsophical Quarterly* 55: 495–503.

Comrie, Bernard 1976. *Aspect.* Cambridge University Press.

Danto, Arthur 1965. "Basic Actions." *American Philosophical Quarterly* 2: 141–8.

Dowdy, David R. 1979. *Word Meaning and Montague Grammar.* Kluwer.

Dretske, Fred 1988. *Explaining Behavior.* MIT Press.

Fara, Michael 2005. "Dispositions and Habituals." *Nous* 39: 43–82.

Handfield, Toby 2008. "Unfinkable Dispositions." *Synthese* 160: 297–308.

Hornsby, Jennifer 1980. *Actions.* Routledge & Kegan Paul.

Johnston, Mark 1992. "How to Speak of the Colors." *Philosophical Studies* 68: 221–63.

Lewis, David 1997. "Finkish Dispositions." *The Philosophical Quarterly* 47: 143–58.

Maier, John 2015. "Dispositions and Ergativity." *The Philosophical Quarterly* 65: 381–95.

Manley, David and John Hawthorne 2005. "Review of Stephen Mumford's *Dispositions*." *Nous* 39: 179–95.

Manley, David and Ryan Wasserman 2008. "On Linking Dispositions and Conditionals." *Mind* 117: 59–84.

Martin, C. B. 1994. "Dispositions and Conditionals." *The Philosophical Quarterly* 44: 1–8.

McCann, Hugh 1975. "Trying, Paralysis, and Volition." *The Review of Metaphysics* 28: 423–42.

McKitrick, Jennifer 2003. "A Case for Extrinsic Dispositions." *Australasian Journal of Philosophy* 81: 155–74.

Molnar, George 1999. "Are Dispositions Reducible?" *The Philosophical Quarterly* 49: 1–17.

Parsons, Terence 1990. *Events in the Semantics of English.* MIT Press.

Setiya, Kieran 2013. "Epistemic Agency: Some Doubts." *Philosophical Issues* 23: 179–98.

Shoemaker, Sydney 1984. *Identity, Cause, and Mind.* Cambridge University Press.

Vetter, Barbara 2015. *Potentiality.* Oxford University Press.

Yablo, Stephen 2010. "Intrinsicness." In *Things*. Oxford University Press, 33–58.

4

Structural Explanation: Garfinkelian Themes

I

Like any science, the social sciences aim to explain things; and some of their explanations of social phenomena are "structural explanations." So what makes structural explanations distinctive? What is structural explanation? And, are structural explanations better than individualistic (non-structural) explanations?

I think we risk talking past each other when we discuss these questions, because I think there are many things we might mean by "structural explanation." On one definition a structural explanation is a special kind of causal explanation, namely a causal explanation that cites some aspect of some structure as a cause of the event being explained. I think it plausible that there are (or can be) explanations that satisfy this definition. For example, religious practice in the United States has declined (in part) because society has become more urban.[1] This satisfies the definition if the factor the explanation cites, namely increased urbanization, (i) is a cause of the decline in religious practice and (ii) is "some aspect of some structure." I think (ii) is plausible: surely the fact that society has become more urban is a fact about how the structure of society has changed. I don't have a definition of "structure" to use to defend (ii), but I think I can recognize central

[1] If this is false, it at least could be true, and its possible truth is all that is needed to establish that there could be explanations satisfying the definition.

examples of facts about structures when I see them, and this is such an example. I also think that (i) is plausible. Increased urbanization passes a standard counterfactual test for causation: if urbanization had not increased, religious practice would not have declined (as much). Note that it is consistent to say both that urbanization is a cause and that there are more basic, non-structural causes of religious decline, and that urbanization is a cause only because these more basic causes are causes.[2]

I admit that this defense is brief: some philosophers may want to challenge (i) or (ii). And even if you accept them, it is puzzling how an event like increased urbanization can cause anything, since it seems much less "oomphy" (for lack of a better word) than an explosion or a billiard ball impact. One might want a theory of causation that addresses this puzzle. But my main interest here is not in giving a thorough defense of the existence of explanations that are structural in the sense that they cite a structural cause, but in isolating a different sense "structural explanation" might have, and arguing that there are explanations meeting this other defintion.

On the sense of "structural explanation" I will be interested in, structural explanations are non-causal explanations. But of course not every non-causal explanation deserves to be called a structural explanation. My procedure for discovering what other conditions structural explanations (in my sense) must meet will be to riff on some ideas Alan Garfinkel presented in his 1981 book *Forms of Explanation.*[3] This book is usually cited for its early expression of the thesis that explanation is contrastive: that "Why did Mary get the A?" can mean "Why did Mary get the A rather than a B?" as well as "Why did Mary rather than John get the A?" But Garfinkel's real

[2] The urbanization example belongs to the first of four types of structural explanation described by Jackson and Pettit (2004, 134). They accept that urbanization is a cause, but not a basic one, so accept that there are structural explanations on this first definition. The question they want to answer is how high-level structural causes can constitute better explanations than low-level individualistic causes.

[3] All references are to this book unless otherwise noted.

interest was in structural explanation.[4] I find the things Garfinkel said about structural explanation suggestive, but somewhat unclear. What I want to do is lay out the definition of structural explanation that they suggested to me. Maybe this definition is the definition Garfinkel had in mind; maybe it's not. Either way, I think the definition is a good one, and it is useful to see how Garfinkel's ideas suggest it.

In outline, an explanation satisfies my definition of "structural explanation" iff it cites a structural fact to explain why a cause had a certain effect. Now explanations (of the kind we're interested in here) are answers to why-questions. In the terminology of answers, part of what makes explanations that meet my definition different from explanations that satisfy the definition I started with is that they are answers to a different kind of question. Structural explanations on the first definition are answers to questions of the form "why did such-and-such happen?," while structural explanations on my definition are answers to questions of the form "why did such-and-such have such-and-such effects?"[5] That's the difference between the two definitions; what they have in common, naturally, is the requirement that an answer to their respective question cites a fact about some structure. At the end of the chapter I will return to the question of whether a structural explanation (on my definition) can be better than an individualistic explanation.

[4] The idea that explanation is contrastive is not original with Garfinkel; Dretske was there first (1972). Garfinkel had other insights in his book, besides his ideas about contrastive explanation and structural explanation. He also made a claim that today we would formulate in terms of proportionality. Stephen Yablo (1992) developed the idea of proportionality in the theory of causation: the idea, roughly, is that the cause should be "enough, but not more than enough" for its effect. When many events "compete" to be a cause of some event, it is the most proportional event that wins. Garfinkel came close to requiring proportionality in explanations. His ideas about proportionality, though, are I think independent of his ideas about structural explanation, so I will not discuss them.

[5] This isn't exactly right; I'll say in more detail what form the questions share in sections IV and V.

II

Garfinkel uses "structure" or "structural" in connection with explanation throughout his book, but not in a single, uniform way. Often he uses "structure" when characterizing the thing being explained. One kind of why-question he wants to answer has the form, why does such-and-such social structure exist? He repeatedly distinguishes questions like this from related ones with which he thinks they can be confused, as when he asks us to "[r]ecall our distinction between the two different questions that can be asked about a social structure: 1. Why is there this structure? (The structural question), and 2. Why, given this structure, does an individual come to occupy a given place in it? (individualistic question)" (123).

This distinction between structural and individualistic why-questions is helpful. But we are after a distinction between structural and individualistic answers to why-questions. Garfinkel thinks there is a connection: he asserts that (at least some) explanations of social structure are different from explanations of other facts. At one point he sketches an explanation of the fact that our society is stratified (this is a fact about the structure of our society), and he claims that "[t]he form of this explanation is important" (122). So Garfinkel thinks that a special object of explanation, namely an aspect of social structure (social stratification), calls, at least sometimes, for a special form of explanation, which we might want to call structural explanation.

What does Garfinkel think this form looks like? Here is how he describes it in one place:

> explanations [like this] are very powerful because they show that the stratification is really explained by the structure, not by the individuals. An explanation which gives us a purely structural answer to the question Why is there social stratification? tells us that such stratification is part of the inherent dynamics of the system, a consequence of the geometry of social relations. (122)

Clearly the form of explanation Garfinkel wants to isolate is some kind of structural explanation, since in it a fact about something's structure is "doing the explaining." The fact that our society is stratified is "really explained by the structure, not the individuals." But which fact about the structure of society is doing the explaining? More importantly, how is the structure doing the explaining? Does the structural fact causally explain why our society is stratified? If so, then the example he's talking about confirms the hypothesis that there are structural explanations on my initial definition, but does not suggest any alternative definition. It would show at most that there are causal explanations that invoke structures (or events involving structures) as causes.

Other things Garfinkel says suggest that he would deny that the structural fact he has in mind causally explains why society is stratified. In outline, Garfinkel's answer to the question of why society is stratified goes like this (he presents it on pages 120–2). First, by "society is stratified" he means that society divides into (i) a small class of people who control vast amounts of wealth and resources, and (ii) a large class of people who control only a small amount of wealth and few resources.[6] Now, says Garfinkel, in a capitalist society like ours social equality, social homogeneity, is unstable. Even small amounts of inequality get magnified into large amounts, since those with a small advantage are able to use that advantage to increase their wealth even more at the expense of those less well-off. This in outline is Garfinkel's explanation of stratification.[7] What is important is that Garfinkel does not explain stratification by citing a past structure-involving event as a cause. He's not pointing to some moment in the past when a tiny bit of inequality crept into our social structure, and saying that current inequality is so vast today because that particular event happened. The core of his explanation is instead the claim that

[6] Garfinkel does not say much about a middle class in his discussion.

[7] I am less interested in whether Garfinkel's explanation is right than in what definition of "structural explanation" it suggests. Probably it is at best over-simplified.

however wealth had been distributed in the past, there would still be a great deal of stratification today. So when Garfinkel says that stratification is "explained by the structure," he cannot mean to be explaining stratification by citing past structure-involving events as causes of that stratification. But then we're back to the question of what structure it is that explains stratification, and the question of how that structure explains stratification.

We get some clues in the rest of the passage I quoted above. A structural answer says that stratification is "part of the inherent dynamics of the system, a consequence of the geometry of social relations." But this cannot be literally true. The dynamics of a system are the rules by which it changes over time. Since stratification is a fact about a society at a time, it cannot literally be "part" of the dynamics of that society.

Chapter 1 of Garfinkel's book contains a section called "Structural Presuppositions." He there discusses a simple example of a class being graded on a curve. He refers to this discussion frequently throughout the book. Those references, as well as the section's title, suggest that understanding the grading-on-a-curve example is the key to understanding what form of explanation Garfinkel thinks structural explanations exhibit. So let's take a close look at that example, and then come back to untangle Garfinkel's claims about structural explanation.

III

Here is what Garfinkel says about the example:

> There is a certain kind of presupposition that arises when the explanations we seek deal with individuals who are related in a larger system. The theory of these presuppositions is the foundation for much of what I am saying in this work.
>
> Let me begin with an example. Suppose that, in a class I am teaching, I announce that the course will be "graded

> on a curve," that is, that I have decided beforehand what the overall distribution of grades is going to be. Let us say, for the sake of the example, that I decide that there will be one A, 24 B's, and 25 C's. The finals come in, and let us say Mary gets the A. She wrote an original and thoughtful final.
>
> Now, if someone asks me why Mary got an A, I would say exactly what I just said: she wrote an original and thoughtful final. Yet this is inadequate as it stands. Suppose *two* people had written finals that were well thought out and original. Would two people have received A's? Not if I am really grading on a curve. Because of this, it is misleading in a certain way to answer why Mary got an A by citing this simple fact about her—that she wrote a good final. It is misleading because it gives the impression that "writing a good final" is sufficient to explain "getting an A." But that is not true. Even if someone writes a good final, they may fail to get an A because someone else has written a *better* one. So it is more accurate to answer the question by pointing to the *relative* fact that Mary wrote the best paper in the class. (41)

Garfinkel calls the fact that there will be one A, twenty-four B's, and so on a "structural presupposition" of the question of why Mary got an A. Now the notion of a structural presupposition is supposed to help us grasp what a structural explanation is. How does it do that? What is the relationship between structural presuppositions and structural explanation?

First let's look at another way Garfinkel characterizes this particular structural presupposition. He characterizes it in terms of the "possibility spaces" of the students in the class, and of the class itself. So first let's see what possibility spaces are.

The possibility space for each student is, in this case, the set of possible grades that a student might receive, namely {A, B, C}. The possibility space for the class, on the other hand, is the set of possible grade assignments the students might collectively receive. This set is too big to list all of its members, but one member, one grade assignment, might start out like this: Mary = A, Bob = B, Harold = B (and so on). Each member of the class's possibility space—each grade assignment—determines an element of each student's possibility space. The (partial) grade assignment I just described, for example, determines the element "A" in Mary's possibility space.

Now if you didn't know that the class was being graded on a curve, you might think that the possibility space for the class was the "product" of all the individual possibility spaces: you might think that every way of selecting one grade from each of the student's possibility spaces determines some member of the class's possibility space. That is, you might think that every combination of individual possibilities is a "global" possibility. But the class *is* being graded on a curve, so this is not true: "certain combinations of individual possibilities are not collectively possible" (44). For example, the combination of individual possibilities in which everyone gets an A is not collectively possible, because it violates the instructor's grading policy. The global possibility space is "truncated."

It is the instructor's grading policy, which Garfinkel originally called a "structural presupposition" and now calls a set of "structural conditions," that rules out as collectively impossible certain combinations of individual possibilities (44). Garfinkel's calling conditions like this "structural" raises a question: why is this the right adjective to use to describe that kind of condition? What makes them structural conditions? How do they differ from "non-structural" conditions? Garfinkel does not elaborate, so here is a stab at an answer. When the possibility space for the class as a whole, the global possibility space, is simply the product of the individual possibility spaces, then in a natural sense the global possibility space is "unstructured": it is just a shapeless bag into which has been thrown every combination of individual possibilities. But when there are restrictions on which combinations of individual possibilities are globally possible the global possibility space does have a kind of shape—a kind of structure. Only certain combinations of individual possibilities "fit" that structure. So what makes the fact that the instructor is grading on a curve a structural condition, on this interpretation anyway, is that it imposes a structure onto the global possibility space.

The idea of a structured possibility space is fascinating, tantalizing. But what exactly is its significance for the theory of explanation? What do structured possibility spaces have to do with structural explanation?

IV

I now want to propose a way to connect the idea of a structured possibility space that emerges from the grading on a curve example to the idea of a special form of explanation, structural explanation, that emerges from the social stratification example.

The first thing to do is to take up the question: in what sense is it impossible for every student to get an A? This is not logically impossible, or physically impossible, or "institutionally" impossible (the school doesn't outlaw such things). Instead, the sense in which every student's getting an A is an impossible state of affairs is that it is an "inaccessible" state of affairs. It is one the class cannot "get to," no matter how they perform on their exams. It helps to think of this in terms of inputs and outputs. The inputs are "raw scores" on student papers. Suppose that the instructor first marks each student paper out of 100; these numbers are the raw scores. Then the instructor applies his curve to the raw scores, to assign grades: the student with the highest raw score gets an A, the student with the next highest score gets a B, and so on. The output of this procedure is an assignment of grades to the students. In these terms, the grade assignment in which every student gets an A is impossible in the sense that, given the instructor's grading policy, no input leads to it as an output.

Now our interest is in explanation, and in what structural explanation might be. How does explanation get into this story about grading policies and possibility spaces? Garfinkel said that the form of explanation he's interested in is one that is instantiated by explanations of structures. And we also noted that in the grading example the global possibility space is structured, in virtue of the fact that certain possibilities are excluded. These together suggest that we examine, as a possible example of a structural explanation, the answer to the question

> Why is the global possibility space structured as it is?

This question can be sharpened, in light of what I've said about it. First, we can be more specific about the way the possibility space is structured:

> Why is it impossible for every student to get an A?

And second, we can be more specific about the kind of impossibility we are talking about:

> Why is it that no possible combination of raw scores results in every student getting an A?

This, I think, is the question we should be focusing on.

And we know the answer: no combination of raw scores results in every student getting an A because the instructor is grading on a curve, a curve on which only the highest raw score gets an A.

The point of this exercise has been to find an answer to a why-question that exhibits a form that deserves to be called "structural explanation." Does this answer have such a form?

Actually I think it does not. I don't think there's anything distinctively structural about the answer. I do think, though, that we are very close. The easiest way to see what the missing piece is will be to look at a different example and see how it differs from this one. I'll do that in the next section.

Although this section has not produced a form of explanation that deserves to be called structural explanation, it has produced a key insight that we should hold on to. The insight is about the kind of why-question that structural explanations (the kind of structural explanations I want to isolate) are answers to. Raw scores stand to grade assignments as causes to effects. The why-question we answered in this section, then, has the abstract form

> Why is it that certain events had, or would have had, certain effects?

The raw scores are the certain events, and each set of raw scores, had the students received it, would have had as an effect a grade assignment in which not every student got an A. The question was why this is so.

It will be a necessary condition on being the kind of structural explanation that I am interested in that the explanation be an answer to a question of this form.

V

Sarah Winchester, the widow of the gun manufacturer, built maybe the strangest mansion ever, in San Jose, California. All but one of the bathrooms are "decoys" built to confuse spirits. There are stairways that go nowhere, straight into the ceiling. Now imagine an even stranger counterpart of the Winchester house, one that contains a completely inaccessible room. Maybe the room has no door. Or maybe it has a door, but there is no way to get to the door: the room could be lofted above another, and its door could open onto thin air twenty feet above the floor.

Now it is impossible to be in that strange room in the same way that it is impossible for all the students to get As. If we think of the set of locations in the house that a person can occupy as a "possibility space," then this possibility space, like the set of grade assignments, is truncated. But, again, it is not intrinsically impossible for all the students to get As. Similarly, it is not intrinsically impossible to be standing in the strange room. What is impossible is gaining access to these situations. No matter how you walk around the house, you won't end up standing in that room, just as no matter what raw scores the students earn, they won't end up all getting As.

To simplify things let's assume that people navigate the house as follows. Before they walk through the front door, they form a collection of intentions about what they will do after they walk in, intentions from which they do not deviate. For example, someone might decide that after they walk in they will walk ten paces, opening any doors they come to (or until they hit a wall), turn left, walk another five paces opening any doors they come to (or until they hit a wall), and so on. These prior intentions are then the "input," and the place in the house where they end up standing when they finish is the "output."

Now we need to look at the answer to the question,

> Why is it that no set of prior intentions, if executed, will result in that person standing in the special room?

We know the answer: no prior intentions have that result because of the way the house is laid out—say, because no path that a human being could traverse leads to the strange room's door.

Now this answer is interestingly different from the answer to the question of why no raw score could earn all the students As. The interesting difference is that the answer to the question about the special room describes the structure of the house. The answer thus cites a (relevant) structural fact. The answer to the question about the grades, by contrast, cited the instructor's grading policy, which seems instead like an individualistic fact, insofar as it is a fact just about the beliefs or intentions of a single person.[8]

We've found the missing ingredient: if we're looking for a variety of structural explanation, we should be looking for explanations that cite structural facts. Let's mix this in with the other condition I've put on the variety of structural explanation that is my target. We can put the definition like this:

> "X because Y" is a structural explanation = df X is the statement that something had (or would have had) a certain effect, or that several things had (or would have had) certain effects; and the fact that Y is a fact about something's structure.

[8] This fact might be in part a fact about the structure of the teacher's brain, which might be a structural fact; but this is not the relevant kind of structural fact.

The explanation in the Winchester house example satisfies this definition.

Both the grading example and the Winchester house example answered a question of the form, "why does no input produce a given output?" But a structural explanation (on my defintion) does not need to answer a question of this form. You also have a structural explanation when you have an answer to a more specific question of the form "why would this input produce that output?" If I actually do enter the house and execute a set of intentions, and I end up standing in the kitchen, then the answer to "Why did executing those intentions result in my standing in the kitchen?" will describe some facts about the structure of the house; and this answer will qualify, according to the definition, as a structural explanation.

VI

I've found the definition of structural explanation I was after. I should stress again that I am proposing this as one possible definition of "structural explanation," not the only possible one.

Do Garfinkel's examples satisfy this definition? I think they do—or, anyway, that there are interpretations of the examples on which they do. Let's start with the social stratification example. Garfinkel wrote that "stratification is really explained by the structure . . . a purely structural answer to the question Why is there social stratification? tells us that such stratification is part of the inherent dynamics of the system, a consequence of the geometry of social relations" (122). To make this a structural explanation we need to ask, not the question Garfinkel explicitly writes down, namely

> Why is there social stratification?,

but instead the question,

> Why is it that almost any "initial state" for our society results in a stratified society?

Garfinkel's answer was something like this: almost any initial state results in a stratified society because our capitalist system makes it easy for small inequalities to get magnified and entrenched. This answer meets the conditions of my definition: the why-question being answered is a question about why certain things would have certain effects, and the answer cites a structural fact about our society, namely its form of economic organization—capitalism.

I've interpreted Garfinkel so that he is answering the question of why social stratification is inevitable, rather than the question of why stratification exists.[9] His actual words, however, are much closer to asking the second question. So how can my interpretation be anything other than a willful misunderstanding?

Actually, Garfinkel words his question differently in different places. He starts the section I've been discussing by asking, "how are we to explain social stratification" (120)? But at the end of the section he says he's explained "why a certain kind of stratification must occur" (122). So Garfinkel's word choice doesn't provide strong evidence either way.

One thing to be said in favor of my interpretation is that it has Garfinkel asking a good question, rather than a bad one. If Garfinkel is right, then the question of why stratification exists is bad because stratification is more or less inevitable, and "Why Q?" is only a good question when there is a "live" alternative to Q. The question of why stratification is inevitable is not bad in this way.

Another thing to be said in favor of my interpretation is that Garfinkel's answer looks more like an answer to the question my interpretation asks. Remember the puzzling passage: "stratification is part of the inherent dynamics of the system, a consequence of the geometry of social relations." Now "the geometry of social relations" looks like a fancy way of describing capitalist economic structures. So just what

[9] I don't mean to suggest that these are the only two questions one might ask about stratification. One might also want to know why stratification persists once it comes into existence, or why stratification is stable (why it is so hard to eliminate it). I think that answers to these questions that cite structural facts will also meet my definition of structural explanation. Thanks here to Sally Haslanger.

is a consequence of those structures? The thing that is in the first instance a consequence of those structures is that small amounts of economic inequality are easily magnified and entrenched—so that inequality is (almost) inevitable. But then his talk about the geometry of social relations is an answer to my question, not to the question of why stratification exists. Similarly, talk of the "inherent dynamics of the system" is talk of the laws or rules that describe how the system (in this case, the distribution of wealth) change over time. The claim that stratification is "part of the inherent dynamics of the system," then, looks like the claim that the laws take pretty much any input to an output with great inequality. Again, this addresses my question more than the question of why stratification exists.[10]

VII

I said at the outset that my definition of structural explanation will make structural explanations different from causal explanations that cite structure-involving causes. Does it do this?

I think it does. A causal explanation explains why something happened by describing its causes. A structural explanation, on my definition, explains why certain things had, or would have, certain effects. Clearly, the thing being explained by a structural explanation isn't the sort of thing that can have causes in the first place. *That executing these intentions would result in your standing in the kitchen* (to return to the Winchester house example) isn't the sort of thing that can have causes. So an answer to the question of why it is so can't cite those causes.[11]

[10] Garfinkel's answer to the question of why stratification exists resembles what Elliott Sober (1983) calls "equilibrium explanations"; in fact it may be an equilibrium explanation. I argue that the core of an equilibrium explanation answers the question of why certain outputs are almost inevitable in (Skow 2016).

[11] The term "causal explanation" is to a certain extent a term of art, and some may prefer to use it in such a way that my structural explanations count as causal explanations. Still, the paradigm case where "A because B" is a causal explanation is one in which (i) the statement going in for "A" describes the

I think this is intuitively true; it also follows from widely accepted views about causation. It is widely accepted that only events can cause, or be caused. But "executing these intentions would result in your standing in the kitchen" does not describe the occurrence of any event. If it did, there would be an answer to "Where and when did it happen?"—but there is not. It also fails to describe the occurrence of any event on the theory of events I defended in chapter 2. I will come back to this point in the next section.

VIII

I think my conception of structural explanation helps make sense of other examples Garfinkel discusses. A few pages after he discusses social stratification Garfinkel writes this:

> We are told...that people have an innate aggressiveness which is used to explain war. But surely the individual trait aggressiveness (whatever it is) does not produce war in every conceivable range of social environments; it must be somewhat sensitive to the range of environments in which it finds itself. We can imagine forms of social organization in which aggressivness would not suface in war, say, by maintaining some kind of social spacing or by channeling the aggression somehow.
>
> This dependence on social structure means that any attempt to explain a sociological trait by appeal to a biological trait will be at best elliptical. There is always a structural presupposition, so we can say once again that the explanatory frame is not
>
> biology → sociology

occurrence of an event E, and (ii) the statement going in for "B" describes one or more causes of E. My structural explanations are far from this paradigm.

> but rather
>
> biology × sociology → sociology.
>
> The effect of suppressing the structural presuppositions is that the resulting statement gives us a false picture of the causalities involved. Consequently, many of these "human nature" explanations are like explaining the existence of restaurants by saying that people have to eat. We can grant that it is human nature that people have to eat, but, we want to ask, why should that necessitate *restaurants*?... All that human nature requires is some way of getting food to people; hence that additional content in the notion of a restaurant is not necessitated by individual nature but rather by the need to satisfy that nature within a very definite context of social relations. (126–7)

I hope that bit about restaurants jumps out at you. Here Garfinkel explicitly focuses on a why-question that is of the kind that structural explanations (on my definition) answer: the question of why it is that human nature's being as it is (this is the "input") resulted in the existence of restaurants (this is the "output"). His answer alludes to a "context of social relations," which, whatever exactly it is, is a structural fact about society. So the restaurant example looks to satisfy my definition of structural explanation.

The aggression/war example satisfies it as well. The example is obviously over-simplified along many dimensions, but we can still draw lessons from it. It may be true that human agressiveness results in wars. The question then arises of why human agressiveness results in wars. Answering this question requires looking at the way our society is structured. It is particular features of that structure that determine how propensities of individual humans, like aggressiveness, result in propensities of the society they make up, like warlikeness.

What about Garfinkel's equation? He wrote that "biology → sociology" is wrong, "biology × sociology → sociology" is right. But it's not entirely clear what these even mean. Here's an interpretation:

yes biological facts produce sociological facts, but this causal relation is itself explained by other sociological facts. Biological facts like inborn tendencies do not produce sociological facts in a vacuum. But "biological facts B produce sociological facts S because of sociological facts Z" is, on my definition, a structural explanation.

There is, however, another possible interpretation of these equations that is worth discussing. Why not take the claim that "biology × sociology → sociology" is the "right" equation to mean, not that sociological facts explain why biological facts produce sociological ones, but that biological facts produce sociological ones only in conjunction with other sociological facts? This interpretation puts the "sociology" on the left-hand side of the arrow on a different level than I have put it on. Let's look at the difference between these two claims:

> Different Levels: sociological facts explain why (biological facts produce sociological facts); vs
>
> Same Level: sociological and biological facts together produce sociological facts.

Here's an argument that only Different Levels is true. It seems to me that, in all the cases we've talked about, the structure that is crucial to the structural explanation is a background condition, not a cause. It is doing what the presence of oxygen is doing when a match is struck in a room containing oxygen and subsequently bursts into flame. Now on one view the presence of oxygen and the striking of the match are joint causes of the lighting of the match.[12] If that view is correct, and the structures are doing what the presence of oxygen is doing, then Same Level is true. But I don't think that view is correct. In chapter 2 I argued for an alternative view: the presence of oxygen is not a cause of the lighting, and in general background conditions are not causes. On my view, only events can be causes, and background conditions are states. Moreover, a state is a background condition iff it explains why some event is a cause of another. This theory of background

[12] This view goes back to John Stuart Mill (1846, 198).

conditions depends on a theory of events. On my theory (see chapters 1 and 2 for more details), an event happens iff something engages in an act, while a state obtains iff something instantiates a property. The presence of oxygen satisfies the condition for being a state, but not the condition for being an event, for by being present in a room the oxygen is not engaging in an act, it instead instantiates the property of being in the room. Since the presence of oxygen explains why the striking of the match caused the lighting of the match, but does not cause that lighting, the analogue of Different Levels is true of it, and the analogue of Same Level is false of it. Now in the Winchester house example, the parts of the house being arranged in such-and-such a way patterns with the presence of oxygen. It satisfies the condition for being a state, but not the condition for being an event. By being arranged in that way the parts of the house are not engaged in an act; instead they instantiate the property of being so-arranged. Since the house's having that structure explains why certain "inputs" result in certain "outputs," but do not cause the outputs, the analogue of Different Levels is true of it, and the analogue of Same Level is false of it. I think the same goes for the other examples: the social relations that underwrite the way people exchange wealth, in the social stratification example; the social structures that underwrite how a society channels individual agression, in the last example.

There's another benefit to regarding the structural facts in structural explanations as background conditions. Some writers on structural explanation, Sally Haslanger for example (Haslanger 2016, 124), say that structural explanations are important because they uncover hidden injustices. Now certainly the social structures that do the explaining in structural explanations are in many cases unjust social structures (we'll see an example in the next section). But in what sense are they hidden? If those structures are background conditions, we have a sense in which they are hidden. Background conditions are by definition "in the background." More specifically, if we want to know why some event occurred, the answer (if we're looking for a causal explanation anyway) should cite causes. The role of background conditions is not to answer the question of why that event happened, but the question of why certain other events caused it. So if we stay at the first level, and only ask why things are the way they are, the answer won't mention background conditions, and as a result we may not even be aware of those background conditions. But when we look for a structural explanation, and ask why the inputs produce the outputs, the answer will mention and thereby uncover the unjust social structures.

IX

In this chapter I've isolated a kind, or form, of explanation that both deserves to be called structural explanation, and makes sense of Garfinkel's claims about structural explanation. Now I want to look at what Sally Haslanger says about structural explanation in her paper "What is a (Social) Structural Explanation?" (Haslanger 2016). Haslanger's paper has lots of interesting ideas and, like this chapter, is partly inspired by Garfinkel's book. It will, therefore, be useful to look at her claims from the perspective that my definition of structural explanation affords.

Haslanger looks at one explanation, which she calls a structural explanation, in detail. Here is the scenario that raises the why-question:

> Imagine a couple, Larry and Lisa, who, we suppose, are equally intelligent, talented, educated, and experienced in the workplace; they have equal power in their relationship, have no prejudices about gender roles, and are equally capable of all domestic tasks and childrearing tasks. Larry and Lisa decide to have children; baby Lulu arrives. They live in a community where decent childcare is beyond their means. Moreover, let's suppose that in this community, as elsewhere, there is a wage gap: women, on average, make only 75% of what men make. Under these conditions, unless Larry and Lisa have special reasons to think that they are unusual in their earning capacities, it is reasonable for Larry

> to work full-time and for Lisa to make adjustments in her work, e.g., to work part-time, to take time off, to take a less demanding job. (122–3)

Haslanger begins by addressing the following why-question: why did Lisa quit her job while Larry kept his? One answer is that Lisa and Larry wanted to choose the option that was best for their family, and believed that her quitting and him not was best. Now this is not a structural explanation—not in my sense, and not in any intuitive sense. Haslanger puts the answer that she says is a structural explanation like this:

> Because Lisa is part of a system that includes Larry, her employer, etc., and given that Larry wasn't going to quit, the employer wasn't going to provide childcare, and she couldn't just leave Lulu home alone, this was her only real option. She might have made a rational choice to quit, but it is inadequate to just point to her choice as if it occurred independently of the workings of the system.
>
> Moreover, to answer [the question] simply in terms of Lisa's choice to stay home with Lulu is inapt. The possibility space for childcare is severely limited in Lisa and Larry's situation; there are structural constraints that make only some options genuine possibilities. Her desires were "canalized" by the circumstances, because she occupies a node in a social structure.... The structural explanation ... shows that women as a group are structurally situated so that it is rational for them to choose options that keep them subordinate. (123–4)

I want to suggest that Haslanger's discussion gains some clarity if we interpret her as describing an explanation that meets my definition of a structural explanation. On this interpretation, it is "inadequate" to cite the desires that made Lisa's choice rational not because citing them is not enough to answer the question of why she quit, but because it is not enough to answer only that question and then end one's inquiry.

Also urgent is the question of why her quitting and Larry's not is the option that would result in the best consequences for their family. To ask this is to ask why an input (a choice) produced an output (a level of welfare for their family), so is the kind of question a structural explanation (on my definition) answers. Many of the factors Haslanger mentions or alludes to in this passage are facts about the structure of our society that are parts of the answer to this question, including the limited childcare options and the fact that men are paid more than equally qualified women. One way in which these are "structural constraints," one way in which these circumstances "canalize" Lisa's desires, is by making it the case that it was quitting that would result in Lisa's desires being best satisfied. And in virtue of making it the case that quitting would have that result, these factors answer the question of why quitting would have that result.

One feature of the Lisa/Larry example that I want to emphasize is that the structural explanation and the individualistic explanation do not compete, in the following sense: it is false that there is one single why-question—the question of why Lisa quit and Larry did not—to which both explanations are answers. Instead, the structural explanation is an answer to a different why-question. The structural explanation answers the question of why Lisa's quitting and Larry's not was their best option.

Now maybe the most natural way to compare explanations is to compare them as answers to a single given question. It follows from what I just said that a structural explanation (as I understand the notion) cannot be compared in this way to an individualistic explanation. It cannot be a better, or a worse, answer to a given question, just because they are not both answers to the same question.

I do not, however, think this means that philosophers and social scientists who say that structural explanations are in some way better than individualistic ones and that it is structural explanations that we should be devoting our energies to finding are making a mistake. For there is another way to compare explanations. We can evaluate and compare questions themselves, not just answers to questions. Some questions are better than others, in the sense that they are better questions to be asking; when one asks better questions, one is more likely to achieve the goals of one's inquiry. And this way of comparing questions carries over to a way of comparing answers: an answer is better iff it is an answer to a better question. Structural explanations can be better than individualistic explanations in this way.

As a result of Lisa's choice she ends up in a subordinate social position. This is an instance of a pattern of injustice, and one goal of our inquiry might be to locate the source of the injustice. If we focus just on answering the question of why she quit it can be hard to even see the injustice. The answer to that question shows her choice to be rational; how can the result of her choice be unjust if the choice was a rational one to make? If, on the other hand, we answer the question of why quitting was her best option, then we find what we are looking for. The structural features of our society that answer this question, including for example pay discrimination against women and limited childcare resources, are the sources of the injustice.

This fact about which question is better to answer corresponds to a fact about what we should do, if we want to fight injustice. Lisa made the right choice in a bad situation. We shouldn't try to change how Lisa makes decisions, we need instead to change the conditions under which she (and others) choose.

X

"Structural explanation" is a term of art. What are the most useful meanings to give it? In this chapter I've proposed one, one that captures an important class of explanations. On my definition a structural explanation is an answer to the question of why one thing caused or would have caused another that cites a structural fact. This definition applies to some examples philosophers have used to illustrate the notion of structural explanation better, I think, than does the initial definition, which applied only to explanations that cite structures as causes.

References

Dretske, Fred 1972. "Contrastive Statements." *The Philosophical Review* 81: 411–37.

Garfinkel, Alan 1981. *Forms of Explanation*. Yale University Press.

Haslanger, Sally 2016. "What is a (Social) Structural Explanation?" *Philosophical Studies* 173: 113–30.

Jackson, Frank and Philip Pettit 2004. "Structural Explanation in Social Theory." In Frank Jackson, Philip Pettit, and Michael Smith, *Mind, Morality, and Explanation*. Oxford University Press, 131–62.

Mill, John Stuart 1846. *A System of Logic*. Harper & Brothers.

Skow, Bradford 2016. *Reasons Why*. Oxford University Press.

Sober, Elliott 1983. "Equilibrium Explanation." *Philosophical Studies* 43: 201–10.

Yablo, Stephen 1992. "Mental Causation." *The Philosophical Review* 101: 245–80.

5

Agent Causation Done Right

I

Here is a common view: when "cause" is used in the most metaphysically basic way, it is events that cause and get caused. Other, metaphysically derivative, uses of "cause" are to be explained in terms of event causation. This includes the use of "cause" on which it is things (that are not events), like people and windows, that cause and get caused, as in the sentence "Jones caused the window to break."

I think that this common view is false. It gets things the wrong way around. My thesis will be that the most basic use of "cause" is the one on which it relates things that are not events, things like Jones and the window. For reasons I will explain below, I will call this kind of causation agent causation.[1]

[1] How widely believed is the common view? One piece of evidence that it is very widely believed is the fact that Paul and Hall's recent comprehensive treatise *Causation: A User's Guide* assumes that it is true—though the authors do express some skepticism about it (2013, 7). An explanation of why the view is widely believed might point to the fact that it is asserted in Donald Davidson's and David Lewis's influential papers on causation (Davidson 1967, Lewis 1986a; only Davidson argues for the thesis). Few of those who hold that event causation is basic try to explain how other kinds of causal claims are made true by event-causal claims. Thomson (2003) is an exception. The view I defend in this essay does not seem to have had many other defenders. E. J. Lowe's defense is the only one I've run across (Lowe 2002, 199–201). I will say something about his main argument below.

You have agent causation when you have a true statement of the form

> X caused Y to Z by Ving,

where terms for things (that are not events) go in for "X" and "Y," a verb phrase with the main verb in its plain form goes in for "Z," and a verb phrase with the main verb in its present-participle form goes in for "Ving": "Jones caused the window to break by throwing a rock" is an instance.[2] Grammatically, the "by Ving" is optional; I will return to the question of whether a statement of "X caused Y to Z" can always be made more explicit by adding "by Ving."

I will take the claim that one use of "cause" is metaphysically more basic than another to mean this:[3] sentences in which "cause" is used in the less basic way can be given "metaphysical truth-conditions" that use "cause" in the more basic way (I use "metaphysical truth-conditions" in more or less the sense Sider gives it in Sider 2011). So if event causation is basic, then an example of a truth-condition for a statement of agent causation in event-causal terms is this:

> Necessarily, "Jones caused the window to break by throwing a rock" is true iff there was a throwing of a rock by Jones, and there was a breaking of the window, and the throwing caused the breaking.

[2] When I discuss causation by omission I will also allow that you have agent causation when you have a true statement of the form "X prevented Y from Zing."

[3] I speak here and throughout of there being two "uses" of "cause," one more metaphysically basic than another; I do not mean this to entail that "cause" is ambiguous, that it really has two meanings. I suspect it does not. Suppose a small bomb goes off in one corner of a courtyard, and in another corner Jones waves red flags with a terrified look on his face; the citizens then panic. The sentence "Jones, and the explosion, caused the panic" is then true. On standard theories of ambiguity (see Zwicky and Sadock 1975), this shows that "caused" has the same meaning in "Jones caused the panic" and "The explosion caused the panic": if it had different meanings, "caused" in the "reduced conjunction" could only have one of those meanings, and would therefore strike us as malformed or in some way deviant, which it does not.

Since I want to reverse the order of analysis, I think this should go the other way. I will say what a general analysis of event-causal claims in agent-causal terms looks like below; as a preview, here is the truth-condition for this particular case:

> Necessarily, "The throwing of the rock caused the breaking of the window" is true iff X caused the window to break by throwing a rock, where X is the agent of the throwing (in this case, X is Jones).

Why do it this way? Because things like people and windows, not to mention quarks and leptons, are metaphysically more basic things than events.[4] Events are ontological parasites, they owe their existence to more fundamental things. But if this is so then any statement about events has a metaphysical truth-condition that does not mention events. Causal statements are just a special case.

That taking agent causation as basic puts things on the right ontological level is not the only argument I have to offer for doing so. My other argument resembles in form a standard kind of argument in metaphysics. When trying to decide which of two notions is more basic (having established that they are not equally basic), we look to our best theory or theories that employ one or another of those notions. We are justified in accepting as more basic whichever notion is employed by our best theory. So, for example, special relativity employs the notion of "spatiotemporal distance" (or "spatiotemporal interval"), while a competing theory, the Neo-Lorentzian theory, employs instead the notions of spatial and temporal distance.[5] To the extent that special relativity is the better theory (this is a common view, but it does not matter for my point whether it is correct), we should regard spatiotemporal distance as more basic than either

[4] I've heard the objection that (i) people are sums of "time-slices," and (ii) time-slices are events, so (iii) some events are more basic then people. Now (i) is certainly controversial, but even if it's right, I reject (ii): time-slices belong to the same ontological category as people and windows, the category of "things"—they're just instantaneous things. They are not events.

[5] (Brown 2005) contains a philosophical history of this debate.

spatial distance or temporal distance. My argument will not quite be that our best theory of causation is one cast in terms of agent causation; I don't have a theory of causation to offer. Instead my argument will be that many problems concerning causation and questions about causation that are difficult to solve or answer when stated in terms of event causation are easy, or easier, to solve and answer when put in terms of agent causation.

II

Events can be causes and effects, but states cannot. I will say more about states in a minute, for now we can just work with an example. When a ball is red, there is a state consisting in its being red that exists as long as it is red. But nothing can cause this state, nor can this state cause anything. My question is: why not? Why can events, but not states, be causes and effects?

Before trying to answer, I should say that while states cannot be causes or effects, they can be intimately related to causes and effects. Take again the state of that ball's being red. While nothing can cause this state, something can cause this state to come into existence (by causing the ball to become red). And while this state cannot cause anything, the coming into existence of this state can cause things, as when the changing of the ball from blue to red causes a child to become excited. But in the first case the effect is an event, and in the second the cause is an event: in both cases the event is the ball's becoming red. Neither is a case where the state itself (the state of the ball being red) was a cause or an effect.

So again: why can't states be causes or effects? Theories of causation that take event causation as basic have no good answer to this question. (At least, I've never seen a theorist of event causation give a good answer.) At best the theory can say that it is part of the nature of the causal relation that it not relate states. But while this answers the question of why in fact states are not causes or effects, it immediately raises the question of why this is in the nature of causation—and the theory has no answer to this question.

If we take agent causation as basic, on the other hand, we can explain why non-basic causation can take the form of event causation, but cannot take the form of state causation.

The explanation starts with a grammatical observation. In any true claim of agent causation, in fact in any such claim that is even possibly true,

(1) X caused Y to Z by Ving,

both the verb phrase that goes in for "Z" and the verb phrase that goes in for "Ving" are non-stative verb phrases. This can be shown just by checking enough substitution instances of (1), but first I need to introduce the distinction between stative and non-stative verbs.

There are several different tests for when a verb is non-stative, but the most useful one here is this one:[6] a verb phrase is non-stative if, but only if, it can be used to complete the form "One thing he/she/it did was..." "Throw the rock" then is a non-stative verb phrase, since "One thing she did was throw the rock" is grammatical. "Be red" on the other hand is stative, as we want, since "One thing the ball did was be red" is not grammatical.

I've claimed that true instances of (1) always have non-stative verb phrases in the place of "Z" and "Ving"; is this really so? Let's check a few instances:

(a) Jones caused the window to break by throwing a rock at it.
(b) Alice caused the cat to become enraged by standing still.
(c) The bomb caused the crowd to panic by exploding.
(d) The room caused the match to light by containing oxygen.
(e) The oxygen in the room caused the match to light by being present.
(f) The light caused the cars to begin moving by being green.
(g) Bobby caused the ball to be red by painting it.

I take it as obvious that each of (a) through (c) could be true; since they also have non-stative verb phrases in for "Z" and "Ving," they are

[6] I discuss another test in chapters 1 and 2.

confirming instances of my thesis. Furthermore, each of (d) through (g) could not be true;[7] since each has a stative verb phrase in for either "Z" or "Ving," they are also confirming instances.

That none of (d) through (g) could be true is maybe less obvious than the claim that each of (a) through (c) could be. It becomes clearer that they couldn't be true if we contrast them with close cousins that could. While it couldn't be that Bobby caused the ball to be red by painting it, he could cause the ball to *become* red. Similarly, while the light could not cause the cars to begin to move by being green, it could cause the cars to begin to move by *turning* green.

The thesis that true agent-causal claims require non-stative verbs can be used to explain why there is event causation but no state causation. The explanation goes by way of the idea that events go with non-stative verbs, and states go with stative verbs. These things go together in the following sense: if you put a non-stative verb phrase (in the active voice) in for "Ved" in "X Ved," then necessarily if X Ved an event (a Ving by X) occurred in virtue of this fact. If instead you put a stative verb phrase in for "Ved," then necessarily if X Ved a state obtained in virtue of this fact.[8]

Now I can explain why there is event causation but not state causation. There is such a non-basic form of causation as event causation because event-causal claims can be given a metaphysical semantics in agent-causal terms on which they are possibly true. In schematic form, the semantics is this:

> Necessarily, "C caused E" is true (where terms referring to events go in for "C" and "E") iff (i) C is a Ving, (ii) E is a Zing, and (iii) the agent of C caused the agent of E to Z by Ving.

As I said, this is schematic; "Ving" and "Zing" need to be replaced in (i) and (ii) by gerundial nouns derived from non-stative verb

[7] Maybe sentences (d) through (g) are ungrammatical; if so, they could not be true. I suspect, however, that they are grammatical.

[8] See chapters 1 and 2 for a more thorough discussion and defense of this theory.

phrases (for example "stabbing" in "there were three stabbings" is the gerundial noun derived from "stab"), and in (iii) "Z" and "Ving" need to be replaced by the appropriate form of the verb phrases themselves (forms in which the main verbs are in the plain form and in present-participle form, respectively).[9] By the theory of events I just sketched, the right-hand side, the truth-condition, can have true instances; so the left-hand side, the event-causal claim, can also have true instances.

To complete the explanation I now need to say why there is not such a non-basic form of causation as state-causation. I've been treating "there is not such a non-basic form of causation as state-causation" as meaning that every statement of state-causation is necessarily false. The reason why this is so is that state-causal sentences cannot be given a metaphysical semantics on which it is possible for them to be true. For if such a metaphysical semantics could be given, it would have to parallel the metaphysical truth-conditions for event-causal claims. But, as I have claimed, the parallel conditions cannot be satisfied. Here is what the conditions look like for an instance of state–state causation (since I am not going for generality, what follows is simplified compared to the event case; the problem is no different for state–event or event–state causation):

> Necessarily, "The state of the light's being red caused the state of the car's being at rest" is true iff the light caused the car to be at rest by being red.

The right-hand side is necessarily false.

[9] By saying this I assume that the language in which these truth-conditions are stated has a non-stative verb for every kind of event, and that every non-stative verb has a corresponding gerundial noun. English certainly isn't like that, anymore than English has a predicate for every property; but this just shows that the language appropriate for giving metaphysical truth-conditions for non-basic truths is not English.

III

Is it really true that states cannot be causes or effects? What about a rock sitting on a table? The table is holding the rock up. Since holding up is a causal notion, surely it follows that the table (or some event or state involving the table) is a cause of some state or event involving the rock. But there are no relevant events. The rock isn't doing anything. It is at rest, and in virtue of this fact a state S obtains, but no event occurs. So if the table is a cause of some state or event involving the rock, it is a cause of S.

But in fact the table is not a cause of S. If the table were a cause of S, the table would be causing the rock to be at rest. But "the table caused the rock to be at rest" is false for the same reason that example (g) is false: "be at rest" is stative.

It is also false that if the table is a cause of some state or event involving the rock, it is a cause of S. This is a case of the causation of an omission. If the table causes a state or event, it causes a negative event, not a state: it causes the rock's failing to fall.

In fact I don't think the table causes either a state or an event. In agent-causal terms, what is happening is that the table is preventing the rock from falling. But on my view "the table prevented the rock from falling" does not entail that the table caused any state or event, not even a negative event. (I'll explain why when I discuss prevention and causation involving omissions in section VI.)

I've had people suggest that the claim that states cannot be causes is inconsistent with the causal theory of perception. They're thinking that a causal theory of perception says

> Someone sees iff the scene before their eyes' being such-and-such causes them to have a visual experience representing the scene as such-and-such.

Since "is such-and-such" is a stative verb phrase, no event corresponds to the truth of "the scene before their eyes is such-and-such"; instead a state corresponds to this truth. So the nominalization "the scene before their eyes' being such-and-such" cannot denote an event; if it denotes either an event or a state, it denotes a state.[10] So this theory as written seems to require that states be causes (assuming as I do that people do in fact see).

Talk of "the scene" is there to achieve generality, but avoiding it by talking about a particular instance of seeing can still make it seem like causal theories of perception require states to be causes. If there is a tree in front of me, then this theory says that I see the tree iff the tree's being in front of me causes me to have a visual experience with the content that tree is in front of me. As before, "is in front of me" is a stative verb phrase.

It would be bad to have to reject all causal theories of perception. Luckily it is not essential to the theory that states be causes. And in fact in at least one well-known defense of the theory, David Lewis's, the theory is not written to require that states be causes. He writes it as

> someone sees if and only if the scene before his eyes causes matching visual experience (1986b, 273; Lewis adds an additional clause by the end of the paper).

So Lewis requires the scene, a thing, to be a cause, not a state involving the scene.

Lewis's theory can easily be written to require an event to be a cause: say that someone sees iff the reflecting of the incident light by the scene (and/or the emitting of light by the scene) in way X causes a matching visual experience. "Reflect" and "emit" are non-stative, so "the reflecting" and "the emitting" denote events.

Not only can the theory be written so as not to require a state to be a cause, it should be. The proper role to assign the state that consists in the scene's being such-and-such is that of a background condition.

[10] Maybe it doesn't denote either kind of thing. I'm not sure myself. So-called "POSSESSIVE -ING" nominalizations are very confusing. Of course if this phrase does not denote a state then it is not a problem for my view. To make things harder on myself I will assume that it does.

On my view (see chapter 2) a background condition to C's causing E explains why C caused E, and that's what this state does: it explains why the light's impinging on the scene causes the scene to reflect light in way X.

IV

"Agent causation" already has a meaning in philosophy, and it is not the meaning I am using it with when I say that the most basic kind of causation is agent causation. On its established meaning, when "X caused Y" is a true statement of agent causation, as it might be in "Jones caused his arm to rise," then X acted intentionally, or acted for a reason, or acted freely.[11] But as I am using "agent causation," none of these things are necessary for the truth of an agent-causal claim. In fact, on my view there are true agent-causal claims, like "The rock broke the window," where the thing doing the causing—here, the rock—is not even capable of acting intentionally, or acting for reasons, or acting freely.

I am willing to risk this confusion, though, because "agent causation" really is the best term for the kind of causation I think is basic. Recall that "X caused Y to Z by Ving" is true only if the verb phrase that goes in for "Ving" is non-stative. And a non-stative verb phrase is one that can be used to complete the form "What X did was..." Now if you have a true completion of this form, then—obviously—X did something. And if X did something, then X acted: for, at least

[11] For traditional theories of agent causation see (Taylor 1966), (Chisholm 1976), (Bishop 1983), or (O'Connor 1995). Despite the dates of these citations, the distinction between agent causation and event causation goes way back: Thomas Reid claimed that in the "proper and strict sense" only an agent can be a cause (he required an agent to be "a being or mind that has power and will to produce the effect"); it is only in a less strict sense "authorized by custom" that we can speak of events as causes (quoted in Van Cleve 2015, 378). Van Cleve comments that "[i]n Reid's day, agent causation was taken for granted and event causation regarded as a Humean innovation of doubtful intelligibility" (377). I do not doubt its intelligibility, only its claim to be basic.

on a very natural sense of "act," to act is just to do something. (Some philosophers do use "act" in a more restrictive sense, so that you act only if you do something for a reason. But this very explanation of the restricted sense proves the existence of the unrestricted sense.) And if X acted, then X was an agent: for an agent is just something that acts. This chain of reasoning establishes that if X caused Y to Z by Ving, then X was (in Ving) an agent. If the rock caused the window to break by hitting it, then the rock was (in hitting the window) an agent. So when you have a true statement using "cause" in (what I think is) the most basic way, the thing that does the causing is an agent; hence, "agent causation" is a good name for this kind of causation.

In "Causation: Omissions" (2003) Judith Thomson worries briefly about how to connect "thing" causation to event causation (she wants to reduce the former to the latter, but that's not relevant to my point here). For one kind of case she sees no problem: she asserts that Suzy caused a breaking of a window (this is "thing–event" causation) iff an event of which Suzy was the agent caused the breaking ("event–event" causation). But she doesn't think this strategy works for all cases. She doen't think it works for "Suzy's rock caused a breaking of a window." Thomson implicitly claims that you can't say for Suzy's rock what she said for Suzy, namely that the rock caused a breaking iff an event of which the rock was the agent caused that breaking, because the rock is not an agent of any events. So what can you say? She makes a suggestion: "Perhaps we can say that for a rock to cause an event y is for there to be an event E such that (i) E is an event in the rock's history, and (ii) E CAUSES y" (Thomson 2003, 85; all-caps "CAUSES" denotes event causation). This isn't ideal, though. When is an event "in the history" of a thing? She doesn't say.

I think that Thomson is driven to treat Suzy and her rock differently because she is working with a thick notion of agency, a notion on which rocks can't be agents—maybe she thinks that agents are things that can act for reasons. All her problems evaporate if we recognize the thin notion of agency I have described. Using the thin notion we can say for the rock what Thomson said for Suzy. We can say that a rock caused an event X iff there is an event of which the rock is an agent that caused X. It's not hard to find such an event for the case of Suzy's rock causing a breaking: the rock hit the window, and since "hit" is non-stative there was an event that was a hitting of the window by the rock; since "the rock" is the subject of "the rock hit the window," the rock is the agent of the hitting; and the hitting caused the breaking.

V

When one event C causes another, there is always an answer to the question of why, or in virtue of what, C was a cause. C has many properties, but only some of them are properties in virtue of which it was a cause.[12] Suppose I mutter in the back of the room, and the lecturer hushes me. The muttering caused the hushing. Now the muttering may have both the property of *being loud* and the property of *being in Japanese*, but still it may have caused the hushing only in virtue of being loud.

Of course we could imagine circumstances where the roles were switched: where the muttering was a cause in virtue of being in Japanese, not in virtue of being loud. We just need to imagine a lecturer who is not bothered by loud muttering but is bothered by Japanese muttering.

It is, however, not true that you can take any property of any cause C of E, and find a circumstance where C is a cause of E in virtue of having that property. People like some things more than others, and Charlie is no exception, except that Charlie is really interested in events. And that muttering—it was Charlie's favorite event. But not only is it false that the muttering caused the hushing in virtue of being Charlie's favorite event, it is necessarily false. There is no scenario in which the muttering is a cause in virtue of being Charlie's favorite event.

So it seems that the properties that can be had by events may be divided into two classes: the properties that can be properties in virtue

[12] I don't know the origin of the idea that an event is a cause in virtue of some but not all of its properties. One place it appears is in (Kim 1999, 17).

of which an event is a cause, and the properties that cannot. What makes the difference?

You might suggest that only the intrinsic properties of an event can be properties in virtue of which an event is a cause. But that's wrong. Extrinsic ones can be properties like that too. Suppose I muttered loudly at noon, when the class started, and that the lecturer is only bothered when people mutter at the beginning of his lecture; after a few minutes he no longer cares. Then the muttering caused the hushing in virtue of happening at noon. But this is not an intrinsic property of the muttering; duplicates of the muttering could happen at any time.

I don't know how to find an answer if we stick to the language of event causation. But I do know how to find one if we use the language of agent causation.

Using the language of agent causation we can state some sufficient conditions for a property of an event to be a property in virtue of which it is a cause. Recall that in agent-causal terms an agent causes something by doing something, and that in event-causal terms it is his doing of that thing that is the event that is the cause. Now if in an agent-causal claim the verb phrase specifying what the agent did contains an adverb, then (i) that adverb corresponds to a property of the cause event, and (ii) that event was a cause in virtue of having that property.

That's very abstract; here is an example. In the agent-causal claim

(2) Jones caused the lecturer to hush him by muttering loudly,

Jones caused something by muttering loudly. In the verb phrase "muttering loudly" the verb is modified by the adverb "loudly." Now this adverb corresponds to a property of events, namely the property of being loud. Furthermore, if (2) is true, then this property is a property in virtue of which the cause-event, the muttering, was a cause:

The muttering caused the hushing in virtue of being loud.

In this case the adjective "loud" as applied to events corresponds to the adverb "loudly." It will be useful to pretend that for every adverb that modifies a non-stative verb there is such an adjective; if "Tly" is schematic for adverbs, "T" is schematic for the corresponding adjective. We can now state our sufficient condition like this:

(SC1) If X caused Y to Z by Ving Tly, then the Ving caused the Zing in virtue of being T.

Again, the condition in (SC1) is stated in the language of agent causation, while the thing it is a condition for is in the language of event causation.

A quick clarification: I want to understand (SC1) so that "Tly" holds the place, not just for adverbs, but for any "adverblike modifier." A prepositional phrase like "in Japanese" is adverblike, so "If Jones caused the lecturer to hush him by muttering in Japanese, then the muttering caused the hushing in virtue of being in Japanese" is an instance of (SC1).[13]

That's one sufficient condition. Here is another. Jones might cause the lecturer to hush him by muttering, but not by muttering in any particular way. That he muttered at all was enough to get the lecturer to hush him. If this is true, then the muttering caused the hushing in virtue of being a muttering: if "Ving" is free of adverbs, then

(SC2) If X caused Y to Z by Ving, then the Ving caused the Zing in virtue of being a Ving.[14]

[13] I take the terminology of adverblike modifiers from (Parsons 1990, 40). He notes that the differences between adverbs and prepositional phrases "are unimportant from the point of view of their semantic function as modifiers" (40), even suggesting that adverbs are disguised prepositional phrases with indefinite objects, so that, for example, "loudly" means "in a loud manner" (53).

[14] Once one sees (SC2) one might wonder whether (SC1) should have a stronger consequent. If X caused Y to Z by Ving Tly, mustn't the Ving be a cause both in virtue of being T and in virtue of being a Ving? If Jones caused the hushing by muttering loudly, wasn't the muttering a cause both because it was loud and because it was a muttering? Not necessarily. Suppose Jones would have caused the hushing if he had fallen over loudly; he needn't have muttered. Then it seems that what mattered was just that the cause was loud, not that it was a muttering. Of course in some situations the lecturer might be especially annoyed by loud mutterings, but not by other loud things. In a

I should be clear: Jones can't mutter without muttering in a particular way. Maybe he did mutter loudly. My claim isn't that he muttered but not in any particular way, but that it was *by muttering* that he caused the lecturer to hush him, not *by muttering loudly* (or softly or in Japanese). If Jones muttered loudly, then *mutter* and *mutter loudly* are distinct things that he did, so it is perfectly possible for only one of them to be the thing by which he caused the lecturer to hush him.

When I say that *mutter* and *mutter loudly* are distinct things that Jones did, I do not mean to distinguish the event that was a muttering from the event that was a loud muttering. Counting events is different from counting things done. When counting events one is counting concrete particulars located in space and time. When counting things done one is counting verb phrases that apply to something (or counting acts, abstract property-like objects, that something is engaged in).

We have two sufficient conditions, stated in the language of agent causation, for a property to be a property in virtue of which an event is a cause. Now the question I am trying to answer is: which properties of events are properties in virtue of which an event may be a cause? Which properties can do that work, and which cannot? I can now give my answer. The properties that can do that work are just the properties that can get to do that work by the mechanisms in (SC1) and (SC2). That is,

> The property of being S is a property in virtue of which an event may be a cause iff "being S" appears in the place of "being T" or "being a Ving" in a possibly-true instance of (SC1) or (SC2).

If this theory is right, then an event can be a cause in virtue of having the property of being loud—a correct result. Similarly for the property of being in Japanese. Moreover, if the theory is right, an

situation like that the stronger claim is true. But the consequent of (SC1) is also true in that situation. The consequent does not rule out the Ving's being a cause in virtue of being a Ving; it just does not "rule it in." I think that (SC1) is the strongest thing that can be said in general.

event cannot be a cause in virtue of having the property of being Charlie's favorite event—another correct result. This is because there is no possibly-true instance of either (SC1) or (SC2) that has "being Charlie's favorite event" in for "being T" or "being a Ving." There is no possibly-true instance like that because there is no instance, even a necessarily false one, like that. The thing going in for "T" in "being T" in (SC1) is supposed to correspond to an adverb that denotes a way of Ving, but *being Charlie's favorite event*, the property of events, does not correspond to a way of doing something. There is, for example, no way of muttering such that when one mutters that way, the (event that is the) muttering is Charlie's favorite event. Similarly, the thing going in for "being a Ving" in (SC2) is supposed to be derived from a non-stative verb phrase "Ving," but "being Charlie's favorite event" is not.

That we can give this answer to the question of which properties of events are properties in virtue of which they may be causes is, I think, evidence that agent-causal talk, that talk of things causing things by doing things (and by doing things in certain ways), is more basic than event causation.

VI

If event causation is basic, then causation of and by omissions requires us to accept the existence of negative events. But if agent causation is basic, causation of and by omissions requires no such thing. Since a metaphysics that requires belief in negative events is worse than one that doesn't (other things being equal), we should take agent causation as basic.

Why must there be negative events, if event causation is basic? Let's focus the argument on the following examples, one of causation by omission, the other of causation of an omission:

> FLOWERS: Jones promises Bloggs that she will water his flowers while he is out of town. She forgets, and does not water the flowers. The flowers die.

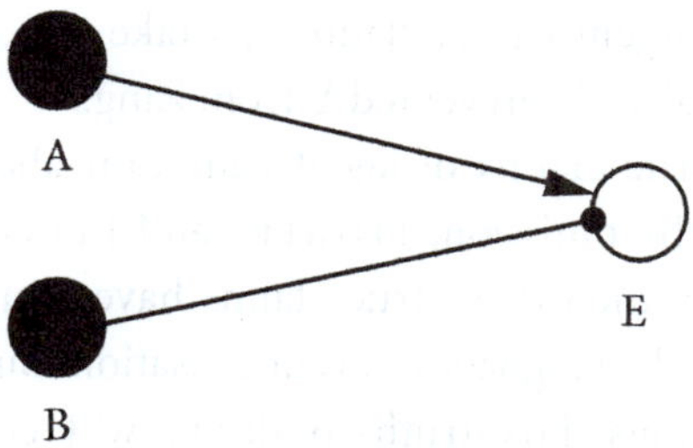

Figure 1

> Prevention: Neuron B sends an inhibitory signal down the axon connecting it to neuron E, just as neuron A sends an excitatory signal to E (see figure 1). E does not fire.

Since Flowers and Prevention are cases of causation involving omissions, they are cases of causation. But I never in my descriptions of those cases say what caused what. If, however, they are indeed cases of causation, we should be able to say what caused what in them. This is easy to do—in agent-causal terms, anyway. In Flowers, Jones caused the flowers to die. In Prevention, neuron B prevented neuron E from firing.

I just said that "B prevented E from firing" is an answer, cast in agent-causal terms, to the question of what caused what in Flowers. But I also said, at the beginning of this chapter, that agent-causal claims have the form "X caused Y to Z (by Ving)." So if what I said at the beginning of this chapter is right then "B prevented E from firing" is not, after all, cast in agent-causal terms. This, however, is not a big problem. There are two possible solutions. One solution is to present metaphysical truth-conditions for "X prevented Y from Zing" in terms of "X caused Y to Z." Then the characterization of Prevention using "prevented" could be replaced by one that used "caused," and the replacement would definitely count as being cast in agent-causal terms. But in fact I doubt that prevention can be analyzed in terms of causation. So I prefer a second solution. I think that it is obvious that agent prevention (as we might call it) is a form of agent causation. So I am going to revise my claim about what agent-causal claims look like. Agent-causal claims can take two forms: the form I first stated, and also "X prevented Y from Zing."

So: there are true claims of agent causation about our cases of causation of and by omission, Flowers and Prevention. If event causation is basic, then these true claims have metaphysical truth-conditions cast in the language of event causation. But if they do have such truth-conditions, those truth-conditions will require us to accept negative events.

For Flowers the sentence that needs truth-conditions in the language of event causation is

(F1) Jones caused the flowers to die.

The most straightforward way to generate truth-conditions for this using event causation yields

(F2) Necessarily, "Jones caused the flowers to die" is true iff some event with Jones as its agent caused the death of the flowers.

Suppose (F2) is true. It's left-hand side is true in Flowers; so then also it's right-hand side is. What event in Flowers, then, witnesses the truth of (F2)'s right-hand side? It must be a negative event. To establish this we need a definition of "negative event." Intuitively, a negative event is one that is in some sense essentially tied to the non-occurrence of some other (kind of) event. Now according to the theory of events that I defended in chapter 2, events go with acts: if someone (or something) is engaged in an act, then a corresponding event is happening. They correspond in this way: if the person is engaged in the act of Ving, then the event is a Ving.[15] A stabbing is

[15] As I explain in chapter 2, this follows from the theory of events I relied on earlier (in section II), which said (in its present progressive version) that "If X is Ving, then an event that is a Ving is happening in virtue of the fact that X is Ving" is true iff V is a non-stative verb phrase. We just need the premise that if someone X is engaged in an act, then something of the form "X is Ving" with V a non-stative verb phrase (in the active voice) is true of them. I discuss the connection between acts and non-stative verbs in more detail later in this chapter. (This argument requires us to work in a language with a non-stative

happening, for example, if someone is engaged in the act of stabbing. My theory also asserts the converse, that if an event is happening, then someone is engaged in a corresponding act. As I will explain below, the converse entails that there are no negative events; I am not assuming its truth here. What matters is the fact that the theory connects events with acts. This fact suggests using acts to define "negative event." Here is my definition:

> E is a negative event iff there is a thing X and an act A such that it is definitional of E that it happens if and only if X does not engage in A (in a contextually determined time and place).

Now in FLOWERS Jones caused the flowers to die because she did not water the flowers. She didn't cause them to die by spraying them with defoliant or setting them on fire. So if there is any event around with Jones as its agent that caused the death of the flowers, it seems it must be an event that is definitionally tied to the fact that Jones did not water the flowers. It seems it must be definitional of that event that it happens iff Jones does not engage in the act of watering the flowers (at the relevant place and time). This is a negative event. So if (F2) is true and its left-hand side is true in FLOWERS, then a negative event exists in FLOWERS. It is natural to regard "Jones' failure to water the flowers" as a name for this event.

Some philosophers have suggested that the event that causes the death of the flowers is the event that exists in virtue of Jones' doing whatever she was doing instead of watering the flowers. Suppose that instead of watering the flowers Jones was doing pushups. Then on this view it was Jones' doing pushups that caused the death of the flowers. This view is usually combined with two more claims: first, that "Jones' failure to water the flowers" denotes, in this case, Jones' doing pushups, so that "Jones' failure to water the flowers caused the death" and "Jones' doing pushups caused the death" are both true, and

verb phrase for every possible act, an assumption that parallels the one I made in footnote 9.)

second, that there is a pragmatic explanation for why only the first is assertible.[16] If this view is right then you can accept (F2) but not negative events, since Jones' doing pushups is not a negative event. It is not definitional of her doing pushups that it happens iff Jones fails to do a certain thing.

This view presupposes that Jones' failure to water the flowers happens (so that "Jones' failure caused the death" is a true event-causal claim), and asserts that the failure is not a negative event. I think this view is false. If Jones' failure exists (and is an event), it is a negative event. I have two arguments.

First, Jones' failure to water the flowers should not be identified with Jones' doing pushups. Doing pushups is, at best, one way to fail to water the flowers. But it is not the only way. Jones still would have failed to water the flowers if she had done situps instead. So if doing pushups is a way to fail, so is doing situps. But in other cases where Ving is a way to Z, and something Zs by Ving, we do not identify the Ving with the Zing. Turning scarlet is a way to turn red, but when a light turns red its turning red is distinct from its turning scarlet. After all, it could be that its turning red caused a bird to peck while its turning scarlet did not. Failures, if there are such things, shouldn't be a special case. So if doing pushups is a way for Jones to fail, but not the only way, then her doing pushups is a distinct event from her failure to water the flowers. And if Jones' failure to water the flowers is not (identical to) a non-negative event like Jones' doing pushups, it must be a negative event.

Second, identifying Jones' failure to water the flowers with the event that occurs in virtue of whatever Jones did instead of water the flowers does not work in all cases. It could be that instead of watering the flowers, Jones did nothing at all. Then there is no event that occurs in virtue of whatever Jones did instead, and so no such event for "Jones' failure to water the flowers" to denote. The only thing left for it to denote is a negative event. And once you accept that it sometimes denotes a negative event, you should say it always does.

[16] Schaffer (2005, 331) holds a view like this.

My argument three paragraphs back that accepting (F2) means accepting negative events relied on a definition of "negative event." Could that definition be challenged? The definition is bad if it entails that every event is a negative event. But it does not entail this. For example, it does not entail that waterings of the flowers are negative events. Suppose Jones waters the flowers, so a watering of the flowers by Jones occurs. Then the following is not true:

(F3) It is definitional of that watering that it happens iff Jones does not engage in the act of "not watering the flowers."

The reason why (F3) is not true is that there is no such act as "the act of not watering the flowers." In general, there is such an act as the act of Ving if and only if it can be true that what X did was V. But it cannot be true that what Jones did was not water the flowers. What can be true is that one thing Jones did not do is water the flowers. But it does not follow from this that one thing Jones did do is not water the flowers. Acts, in this respect, are different from properties. Properties have negations: the negation of P is a property ¬P such that it is definitional of ¬P that something instantiates ¬P iff it does not instantiate P. But acts do not have negations. For an act A, there is not an act ¬A such that it is definitional of ¬A that something is engaged in ¬A iff it is not engaged in A.

Why don't acts have negations? Some opponents of negative acts argue that if there were such things, there would be questions about them we would have no idea how to answer.[17] For example, if there is such an act as the act of "not watering the flowers," then there is the question of how many times in the last week Jones performed that act, and also the question of where those performances took place. Rather than accept the existence of these embarrassing questions, we should deny the existence of negative acts.

[17] See (Galton 1984, 27) and (Paul and Hall 2013, 179). These authors are actually arguing against negative events, not negative acts, but the dialectic is the same.

This line of opposition to negative acts concedes that the idea of a negative act is at least intelligible. I oppose them at an earlier point; I deny that the idea is intelligible.

I earlier said that non-stative verb phrases go with events, and stative verb phrases go with states. It is also true that acts go with non-stative verb phrases and properties go with stative verb phrases. They go together in the following sense: to any stative verb phrase V there is a corresponding property, the property of Ving. So to "X is red" there corresponds the property of being red. But non-stative verb phrases do not have corresponding properties; they instead have corresponding acts. To a non-stative verb phrase V there corresponds the act of Ving. So to "X is running" there is no corresponding property. If there were it would be called "the property of running"—but this expression cannot name a property, or anything else, since it is grammatically ill-formed. Instead, to "X is running" there corresponds the act of running.[18] (The act corresponds to the verb phrase in the sense that it is definitional of the act of running that someone is running iff he is engaged in that act.)

Now I just said that every non-stative verb phrase has a corresponding act, but that isn't quite true. What is true is that every *positive* non-stative verb phrase has a corresponding act. Negative non-stative verb phrases do not. A negative non-stative verb phrase (in the progressive aspect) is one with "not" between "is" and the present-participle form of the verb. "X is not running" is an example. That there is no act corresponding to negative verb phrases like this is made clear by the grammar: if there were, it would be called "the act of not running," but this description does not denote anything, because it is ill-formed.

I've focused on FLOWERS for my argument that if event causation is basic, causation involving omissions requires negative events. But PREVENTION would serve just as well. The event-causal truth-condition for

(P1) B prevented E from firing

[18] I develop this argument in more detail in chapter 2.

has to have the form

(P2) Necessarily, "B prevented E from firing" is true iff some event with B as its agent (in this case, B's firing) caused...

And the most natural way to complete (P2) is to use a term denoting an event of which it is definitional that it happens iff E does not engage in the act of firing.

The question is whether there are any alternatives to (F2) and (P2), any truth-conditions cast in terms of event causation that do not require the existence of negative events in FLOWERS and PREVENTION. The answer is no—or, at least, I've never seen any proposed, and can't think of any myself.

It is important to distinguish the question of whether (F1) and (P1) have metaphysical truth-conditions cast in terms of event causation that do not require the existence of negative events from the question of whether (F1) and (P1) have some metaphysical truth-conditions or other that do not require the existence of negative events. I suspect that the answer to this second question is yes, but even if it is, it does not follow that those truth-conditions will be cast in terms of event causation. For example one might suggest the following truth-condition for (F1):

(F4) Necessarily, "Jones caused the flowers to die" is true iff either (i) for some act A, Jones caused the flowers to die by engaging in A (this covers the case where he did not cause the death by omission); or (ii) there is some act A that Jones did not engage in such that had Jones engaged in it, the flowers wouldn't have died.[19]

[19] I'm not putting this forward as correct. Clause (ii) is too simple. Suppose a powerful computer is monitoring Jones' behavior, programmed to kill the flowers itself if Jones engages in any acts that would otherwise save the flowers. Jones does not engage in any such acts; the flowers die. Then (ii) is false (as is (i)), even though Jones did cause the flowers to die. I'm presenting (F4) just as an example of something that might be on the right track, at least if you like counterfactual theories of causation.

The right-hand side of (F4) does not require the existence of any negative events to be true. Of course, the right-hand side does not speak explicitly about events at all, so it does not require the existence of any events whatsoever to be true. But even if we re-write (F4) to speak explicitly of events, the right-hand side that results will not require the existence of any negative events:

(F5) Necessarily, "Jones caused the flowers to die" is true iff either (i) some event with Jones as its agent caused the death of the flowers; or (ii) there is an event-type such that had an event of that type with Jones as its agent occurred (for example, a watering of the flowers by Jones), the death of the flowers would not have occurred.

But neither (F4) nor (F5) is what we are looking for: neither provides truth-conditions for (F1) in terms of event causation. Claim (F5) of course provides truth-conditions in terms of events, but that is not the same thing. Clause (ii) of (F5) doesn't use the notion of cause at all; it would have to, for it to be a truth-condition case in terms of event causation.

To put this another way: if event causation is basic, if the basic causal locution is "(event) C caused (event) E," then since Flowers and Prevention are examples of causation, some event-causal claim is true in them. But, I've argued, if an event-causal claim is true in them, it entails the existence of negative events.[20]

So: if event causation is basic, and there is such a thing as causation by and of omissions, then you've got to believe in negative events. Not so if agent causation is basic. Then we can just stick with the

[20] Paul and Hall (2013, 213) say things that imply that they think that "The omission of a watering caused the death of the flowers" is the most basic causal claim true in Flowers. They don't regard "The omission of a watering" as a name of an event, though. Presumably then it names the fact that Jones didn't water the flowers (or the fact that no one did). So they seem implicitly to opt for taking the basic causal locution to relate facts (or to relate both facts and events). I am about to discuss the idea that the most basic use of "cause" relates facts.

characterizations of FLOWERS and of PREVENTION cast in terms of agent causation: neither "Jones caused the flowers to die" nor "B prevented E from firing" entails the existence of any negative events.

This isn't quite the whole story. As I've said: if agent causation is basic, then event-causal claims have metaphysical truth-conditions cast in the language of agent causation. So "The hitting of the window by the rock caused the breaking of the window" has the truth-condition "The rock caused the window to break by hitting it." One might pointedly ask whether this works "in the other direction." If every event-causal claim has a metaphysical truth-condition cast in the language of agent causation, shouldn't it also be true that every agent-causal claim is the metaphysical truth-condition for some statement cast in the language of event causation? If so, then "Jones caused the flowers to die, but not by doing anything" in particular is the metaphysical truth-condition for an event-causal claim. And that event-causal claim must be something like "Some negative event with Jones as its agent caused the death of the flowers." But then we are stuck with negative events, even if agent causation is basic.

In reply I reject the claim that things work in the other direction. Since agent causation is basic, every event-causal claim has a metaphysical truth-condition in agent-causal terms: that's right. But the converse is false. Not every agent-causal claim is the metaphysical truth-condition for some event-causal claim. Why should it be? The claim that a certain kind of vocabulary is non-basic entails that statements couched in that vocabulary have more basic truth-conditions. But from the claim that a certain kind of vocabulary is basic nothing follows about whether that vocabulary appears in any truth-conditions for anything else, or about which kinds of claims containing that vocabulary appear in truth-conditions.

I want to emphasize that my argument in this section has not been that we can solve all the problems of causation by and of omission if we cast our theories in agent-causal rather than event-causal terms. One standard problem is the problem of "infection by the normative": Jones caused the flowers to die, the Queen of England did not; there is a difference in causation even though there is no difference in watering-behavior. Neither watered the flowers. One proposal is that the fact that Jones promised to water the flowers is what makes the difference (see McGrath 2005). Does it really? And if so, how could that be? I don't have answers to these questions, nor does thinking in terms of agent causation seem to be much help with them.[21]

An argument essentially like the one I've given against taking event causation to be basic has been around for a while, but usually the alternative offered is taking "fact causation" as basic (see for example Mellor 2004). A standard version of this view says that the basic causal locution is

(Fact) X is a cause of (fact) Y,

where names, or descriptions, of facts go in for "X" and "Y." On this theory, when Suzy breaks a window by throwing a rock, the most basic causal truth is that the fact that Suzy threw a rock is a cause of the fact that the window broke. I can give two reasons why it is better to solve the problem of omissions by taking agent causation as basic than to solve it by taking fact causation as basic.

First, facts just cannot be causes, or effects, in any ordinary sense of "cause." The fact that Suzy threw a rock just cannot cause anything, in

[21] Does taking agent causation as basic help with cases of "extreme causation by omission"? These are cases where, intuitively speaking, the cause of something is not something's failing to do something, but instead is something's failing to exist in the first place. Paul and Hall give this example (2013, 181): some rabbits are introduced into an ecosystem that is free of predators, and the rabbit population increases quickly. If you are okay with negative events you can say that the lack of predators (presumably a negative event) caused the increase in population. My strategy for avoiding negative events has been to couch causal claims in agent-causal terms. So what agent-causal claim is true in this scenario? Not "the foxes caused the rabbits to multiply because they didn't eat them"; there are no foxes. Not "all the non-rabbits in the ecosystem caused the rabbits to multiply because they didn't eat them"; it could be that there are no non-rabbits.

This search has not been exhaustive; there is a true agent-causal claim about this scenario. It uses the notion of prevention: nothing (no thing) prevented the rabbits from multiplying.

any ordinary sense of cause. Facts just aren't the right sorts of things to be causes.

I am not the first to make this complaint. I wanted to say something about its origin, but the complaint's popularity has made its history elusive. It has been made often enough and for long enough that even back in 1988 Jonathan Bennett discussed it without citing any complainer in particular (1988, 22). The complaint, as it appears in Bennett's discussion, may be interpreted as the following argument: (i) something can be a cause only if it can pull something or shove something or bend something or…; (ii) facts cannot do any of those things; so facts cannot be causes.

I think this is a good argument. Bennett, a fan of fact causation, rejected premise (i), saying it

> rests on the mistaken assumption that causal statements must report relations between shovers and forcers. I grant that facts cannot behave like elbows in the ribs, but we know what items do play that role—namely elbows. In our world the pushing and shoving and forcing are done by things—elementary particles and aggregates of them—and not by any relata of the causal relation. (22)

But it is false that the things that push and shove cannot be the relata of the causal relation. A person can shove you, and a person can be a relatum of the causal relation, as Jones is in "Jones caused you to cry." (I take it that something can be a relatum of the causal relation if it can be denoted by the subject of a true sentence of the form "X caused Y.") Similarly, an event can bend things, and an event can be a relatum of the causal relation. A big enough explosion can bend a bicycle around a pole.[22] That things and events meet the necessary

[22] Bennett asked himself "do you deny that when an explosion causes a fire, the explosion…pushes things around?" and answered "Yes, I do," offering this explanation of why people might be tempted to disagree: "The idea that the pushing is done not by the molecules but by the explosion is just the afterglow of ignorance about what an explosion is" (23). But this is a false dilemma: the explosion and the molecules do the pushing—the

condition in (i) is evidence that it is true; that facts do not suggests that facts cannot be causes.

Related to this argument are arguments with more general premises. I endorse these arguments too: facts are abstract, and causes must be concrete; facts have no locations in space and time, but causes must have such locations (note that both rocks and events like throwings of rocks are concrete and have locations).

I said that facts cannot be causes "in any ordinary sense of cause" because it is open to a philosopher to say that the most basic causal locution uses "cause" with a technical meaning that it does not have in English, and that on that meaning "fact X is a cause of fact Y" *can* be true. Philosophers are of course free to give words new meanings, meanings unavailable in English, especially if one can build a better theory using words with those new meanings. Scientists do it all the time. But when you do that you have to give your audience some handle on what the new meaning of your word is; and the fans of fact causation have not done that.

This objection does not apply to every theory of fact causation. One could say that the basic locution is not "(Fact) X is a cause of (fact) Y," but instead "That X caused it to be the case that Y," where (declarative) clauses that express facts go in for "X" and "Y," not names or descriptions of facts. "That Suzy threw a rock caused it to be the case that the window broke" is perfectly ordinary English.

My second objection applies even to these theories. Taking fact causation as basic runs into some of the same problems that taking event causation as basic does, problems that we have already seen taking agent causation as basic can solve. If we allow that facts are the right kind of thing to be causes, still not just any fact can be a cause. For example, this couldn't be true: that the light was green caused it to be the case that the car was moving. On the other hand, this could be true: that I struck the match caused it to be the case that the match lit. What's the difference? The difference is that the fact that the light was

explosion pushes things around in virtue of the fact that certain molecules push things around.

green is a fact expressed with a stative verb, while the fact that I struck the match is a fact expressed with a non-stative verb. But this answer just raises another question: what does the stative/non-stative contrast have to do with the cause/non-cause contrast? If fact causation is basic, it is just a brute fact that only clauses with non-stative verbs can fill the blanks in "that...caused it to be the case that..." to yield a truth. If agent causation is basic, on the other hand, we can parallel what I said about state vs event causation: the reason why only fact-causal claims that use non-stative verbs can be true is that only they have possibly-true metaphysical truth-conditions cast in agent-causal terms.

The main argument of this section has been premised on the claim that it would be bad to believe in negative events. Why is this supposed to be bad? If there are negative events then we should be able to count them; but we can't. If there is a kind of event that consists in Jones' not watering the flowers, how many events of that kind happen in FLOWERS? One? Five? Seventy-two? There is just no saying.

Another argument against negative events is more internal to my own views. I hold that if there were a negative event that consisted in Jones' not watering the flowers, there should be a description of it derivable from the sentence "Jones did not water the flowers," as the description "a watering of the flowers by Jones" is derivable from "Jones watered the flowers." But there is not. What could it be? "A not watering of the flowers by Jones"? That is nonsense.

At this point someone might raise the issue of failure. If Jones didn't water the flowers, then he failed to water the flowers. And if he failed to water the flowers, wasn't there a failure to water the flowers by Jones? Failures are certainly events,[23] if they exist; and, as I argued earlier, if they exist, they are certainly negative events.

I cannot simply deny the existence of failures, for their existence appears to follow from my own theory of events. For "fail" is a non-stative verb, and I've said that whenever "X Ved" is true and what goes in for "Ved" is non-stative, there was a Ving by X, and Vings are events.

[23] Or at least there is one use of "failure" on which it purports to denote events. I don't mean to deny uses like "That guy is a complete failure."

I've thought a lot about failure while working on this section. My considered view is that I should revise my claim: it is not true that whenever "X Ved" is true and what goes in for "Ved" is non-stative and active, there was a Ving by X. This is not true, in particular, when what goes in for "Ved" is a verb phrase headed by "fail." So even when Jones failed to water the flowers, there was no failing to water the flowers by Jones.

Why is "fail" an exception? Because the behavior of this verb is exceptional. Here is a feature of the verb: consider a case where its complement is a non-finite active clause, as in "Jones failed to water the flowers." Consider also the sentence that results from (i) making the complement clause passive, and (ii) swapping the subject of "fail" with the object of "water"; this is the sentence "The flowers failed to be watered by Jones." Now note that these two sentences are analytically equivalent: "Jones failed to water the flowers iff the flowers failed to be watered by Jones" is analytic. But then if there are such things as failures they behave very differently from more garden-variety events like waterings. For if there are failures, and if Jones failed to water the flowers, then there is a failure that had Jones as its agent (since Jones did the failing); but since "Jones failed to water the flowers" is analytically equivalent to "the flowers failed to be watered by Jones," it follows just from the meaning of what has been said that there is also a failure that had the flowers as its agents (since according to this second sentence it is the flowers that do the failing). Surely that's nuts!

This doesn't happen with "water": true, "Jones watered the flowers" is analytically equivalent to "The flowers were watered by Jones," but this second sentence is not an instance of "X Ved" with an active non-stative verb phrase taking the place of "Ved" (recall that I restricted true instances to those with active verb phrases); the verb phrase is instead passive. (Although "be watered by Jones" in "The flowers failed to be watered by Jones" is passive, the verb phrase it is a part of, the one headed by "fail," is active.)

So there are no failures (no events that are failures). "Jones failed to water the flowers" just means that he did not water the flowers (which itself means the same as "the flowers were not watered by Jones"), and the first sentence entails the existence of only the events the second sentences do, namely none.

Is "fail" an isolated example? Is it the only exception to the claim that "X Ved iff there was a Ving by X" is true iff V is active and non-stative? I suspect not. Linguists take the fact that "Jones failed to water the flowers" is equivalent to "The flowers failed to be watered by Jones" to show that "Jones" in "Jones failed . . . " is a "raised subject"; roughly this means that "Jones" belongs semantically with the complement clause "to water the flowers," even though it is syntactically the subject of "fail" (see for example Huddleston 2002, 1195). (The syntactic role of "Jones" better matches its semantic role in "It failed to be the case that Jones watered the flowers," which is analytically equivalent to both "Jones failed to water the flowers" and "The flowers failed to be watered by Jones.") Perhaps it is true in general that non-stative verbs that have raised subjects are exceptions. After all, the following thesis is plausible: "X Ved" entails "There was a Ving by X" only if "X" in "X Ved" is semantically related to the main verb in the verb phrase "Ved." But a raised subject is not semantically related to the main verb.

But the fact that Jones did not water the flowers must be related to the flower's death somehow, mustn't it? And if it is, mustn't there be an event of his failure, an event that can then be a cause of the death? I say yes to the first and no to the second. The fact that Jones did not water the flowers can be related to the death, not by grounding the existence of a cause of that death, but by explaining why Jones caused the death. That is, the fact appears in the proper place when we say: Jones caused the flowers to die, because she did not water them.

VII

About Flowers I asserted that Jones caused the flowers to die, but I did not assert any instance of "Jones caused the flowers to die by Zing." That was of course no accident: if Jones had caused the flowers to die by Zing, then she would have caused the flowers to die by doing something, but the key fact about the case is that Jones did not do anything to cause the flowers to die; that is why it is a case of causation by omission.

This returns us to the question, which I raised and set aside at the beginning of this chapter, of whether "X caused Y to Z" can always be made more complete by adding something of the form "by Zing." I believe that causation by omission shows clearly that the answer to this question is no.

Are cases of causation by omission the only cases in which one thing causes another to do something, but not by doing something itself? If there are other such cases, they are also problems for the thesis that event causation is basic.

Structurally, the question resembles the question of whether anyone ever performs a basic action. If I do one thing by doing another, then in doing the first thing I am not performing a basic action. So if I signal a cab by waving my hand, then in signaling the cab I am not performing a basic action. I perform a basic action when I do something but do not do it by doing something else (the basic/non-basic action distinction was introduced in Danto 1965). The question of whether anyone ever does something without doing it by doing something (else) is clearly close to the question of whether anyone ever causes something without causing it by doing something.

In fact, you might think that a yes answer to the first question entails a yes answer to the second. Many philosophers think that I can, say, clench my fist as a basic action. Suppose they're right. Now "X clenched his fist" entails "X caused his fist to become clenched," where "clenched" in the first sentence is a transitive verb and in the second is an adjective.[24] So if I clenched my fist but didn't clench it by doing anything (else), doesn't it follow that I caused my fist to become clenched but not by doing anything? It might be tempting to say no, that I caused my fist to become clenched *by clenching it*, but I don't think that could be right. To clench one's fist is (at least in part) to

[24] This is not an isolated case; for many "causative verbs" there is an etymologically related adjective such that "X Ved Y" entails "X caused Y to become ADJ." See chapter 6 of (Parsons 1990).

cause it to become clenched; *causing one's fist to become clenched* is part of (maybe all of) *clenching one's fist.*[25] But if to do A is in part to do B, then one does not do B by doing A. You do not perform the part by performing the whole.

If it is false that I caused my fist to become clenched by doing something, we have another agent-causal claim that lacks metaphysical truth-conditions in event-causal terms.[26]

The problem with this argument, I think, is at the first step: the claim that one can clench one's fist as a basic action. There is a simple conflation going on: we must distinguish basic actions from basic intentional actions. A case of basic intentional action is when someone does something intentionally, but does not do it by intentionally doing something else. It is certainly plausible that I can clench my fist intentionally without clenching it intentionally by doing something else intentionally; so it is plausible that I can clench my fist as a basic intentional action. It does not follow that I can clench my fist as a basic action (full-stop). Surely even when I clench my fist in the normal way, I do it by contracting certain muscles in my arm? Maybe I do not always know that I am contracting those muscles, but that shows at most that contracting those muscles is not something I do intentionally, not that it is not something I do.[27]

Maybe clenching one's fist is not, then, a basic action. So what? Surely there are basic actions? If I do X by (in part) doing Y, and do Y (in part) by doing Z, surely this chain must come to an end, and whatever act is at the end is a basic act. This is the standard argument for the existence of basic actions (it appears in Danto 1965). I am not sure it is good. When I walk across the room I walk across it in part

[25] If I order my henchman to break your window, I have certainly caused your window to break. Have I broken your window? Maybe not. If not, then *breaking your window* is *causing your window to break* plus some other condition. Even if that's right, the second thing is still part of the first.

[26] This is, in its essentials, Lowe's main argument that agent causation cannot be reduced to event causation, in (Lowe 2002, 200).

[27] The claim that if you are doing something intentionally you know you are doing it traces back to Anscombe (1957).

by walking across the first half, and I do that in part by walking across the first quarter, and this chain does not come to an end.[28]

But I think this issue can be bypassed. Even if we never perform basic actions, other things certainly do. Consider any electron in the universe. Assuming for simplicity that Newton was right about gravity, that electron is pulling, gravitationally, on the Earth right now, and doing so as a basic action. It is not pulling on the Earth by doing something else. ("Exerting a force on the Earth" is just a fancy phrase for "pulling on the Earth.")

Interestingly, examples of basic actions like this are not examples of something causing something but not by doing something. The problem is that "pull" is not a causative verb. "Pull" is different from "clench." While to clench one's fist is (at least in part) to cause it to become clenched, it is false that to pull on something is to cause it to become X (for any X).

Of course the electron can cause things *by pulling on the Earth*; it can cause the Earth to accelerate by pulling on it. But that's not the kind of claim we're looking for.

What we need, to have an example of an agent-causal claim that couldn't have truth-conditions couched in terms of event causation, is an example of something that can perform a basic action that can be described with a causative verb. Now I haven't found an actual example of this (though maybe I would find some if I looked harder), but surely there are possible examples of it? Couldn't there be a god who could move material bodies about directly, not by doing anything else? If there could, then it would thereby be causing bodies to move but not by doing anything else.

VIII

The simple counterfactual theory of event causation,

(C) C is a cause of E iff had C not happened, E would not have happened,

[28] Thompson (2008, 107–8) tries to show that there are no basic intentional actions using a similar argument; for a reply see (Setiya 2012, 288).

stands refuted by a variety of examples. Most challenging perhaps are cases of preemption, when a backup cause is poised to cause E should C fail to do so. Justly famous is the example of Suzy and Billy: Suzy throws a rock and, by throwing it, breaks a window; Billy is ready to throw if Suzy does not. Suzy's throw causes the breaking, even though had her throw not happened, the breaking still would have happened, for it would have been caused in that case by a throw by Billy.

It is often assumed that the simple counterfactual theory is close enough to the truth to be useful as long as "nothing funny is going on," where the main kind of funny-business to worry about is preemption. If you can be reasonably confident that there are no backup causes ready to step in, then you can be reasonably confident, at least in this case, that C is a cause of E iff had C not happened, E wouldn't have happened.

But even when there are no backup causes the simple theory can fail. Suppose that Suzy is firmly committed to breaking the window one way or another, but has not yet decided on the best means. Should she throw her rock at the window? Or launch the rock with her slingshot? Her aim is better when throwing, but the slingshot generates more force... she decides to throw. Her throwing of the rock causes the breaking of the window. Now it is plausible in this case that had she not thrown the rock, she would have launched it with the slingshot. (If you doubt that this is plausible in this case, certainly there are circumstances in which the counterfactual is true; imagine them to obtain.) So if her throwing hadn't happened, a launching by her would have happened; and if her throwing hadn't happened and a launching by her had happened, the breaking still would have happened (we may suppose that her aim with the slingshot is good enough). It follows from these two counterfactuals that if her throwing hadn't happened, the breaking still would have happened, and we have a counterexample to the simple counterfactual theory.

It is common for philosophers to use a simplified version of Lewis's and Stalnaker's semantics for counterfactuals when reasoning with counterfactuals, a simplified version that says that "If it had been that A, it would have been that B" is true at a world W iff B is true at the world closest to W at which A is true.[29] In these terms, the throwing/slingshot example is one in which the closest world in which the cause—the throw—fails to occur is one in which another event—a launching—occurs in its place that then goes on to cause what the throw actually causes.

This problem for the simple counterfactual theory is a problem for pretty much any counterfactual theory of causation—pretty much all of them are equivalent to the simple theory when there is no preemption (or, more generally, any overdetermination) to worry about.[30] So all these counterfactual theories need a solution to this problem. I'll focus on how it might be solved for the simple theory.

David Lewis noticed the problem and said this about it:

> When asked to suppose counterfactually that C does not occur, we don't really look for the very closest possible world where C's conditions of occurrence are not quite satisfied. Rather, we imagine that C is completely and cleanly excised from history, leaving behind no fragment or approximation of itself. (2004, 90)

I'll interpret Lewis to have amended (C) by strengthening the antecedent:

(C2) C is a cause of E iff had C not happened and no fragment or approximation of C had happened either, E would not have happened.

(We could also interpret him as specifying in more detail the closeness relation we are to use when interpreting (C). I'm using my interpretation only because it is easier to work with.)

[29] Lewis 1973; Stalnaker 1968; the simplified version is closer to Stalnaker's semantics than Lewis's.

[30] It is certainly a problem for counterfactual theories that say that causation is a matter of "conditional dependence." Theories of the form say that C is a cause of E iff had C not happened but F had still obtained, E would not have happened. For examples of such theories see Yablo (2010) and Hitchcock (2001).

But what does a world look like in which C has been "completely and cleanly excised," when no fragment or approximation of C happens in its place? When Paul and Hall raise these questions you can almost see them tearing their hair out over them:

> What exactly does such "complete and clean" excision consist in? Removal of the event by some sort of metaphysical scalpel? Leaving behind... what? The Void? (2013, 51)

In a world where Suzy launches the rock with a slingshot, hasn't the throw been completely and cleanly excised—just also replaced with another event, a launching? Maybe by "completely and cleanly excised" Lewis means "completely and cleanly excised and not replaced by any other similar event." But then we need to know what "similar" means here. If Suzy had kicked the rock, or shot a bullet from a gun (and so not used the rock at all), would the events that replace the throwing be different enough?

It will not do to interpret Lewis to mean "completely and cleanly excised and not replaced by any other event." Suzy could cause Billy to startle by making a face at him; but "If the face-making had not happened and nothing had happened in its place, the startling would not have happened" is false. Billy would still have been startled, if instead of making a face Suzy had suddenly vanished and left behind THE VOID. The cause event—the throwing, the face-making—has to be replaced by something; the question is what.

Drawing on (Maudlin 2004), Paul and Hall propose an alternative to (C2) for which the questions about similarity and approximation do not arise. Their alternative invokes a distinction between a thing's "default" state and its "deviant" states:[31]

> for any given event, we work with an antecedently understood notion of a *default state* for the region in which

[31] Maudlin draws the default/deviant (he actually uses "inertial/deviant") distinction, but does not use it to defend a counterfactual theory of causation; indeed his article is devoted to attacking such theories.

> the event occurs, or for the physical system or systems to which it pertains. Conceiving of the event as one among various possible *deviations* from that default state, we answer the question, "What would have happened, had that event not occurred?" by returning the relevant region or system to its default state, holding the state of everything else fixed. (51)

If we construe this as proposing an alternative to (C2), it amounts to the following:

(C3) C is a cause of E iff had C not happened, and the "system to which C pertains" had been in its default state, then E wouldn't have happened.[32]

If we assume that Suzy's default state is a state where she and all her limbs are at rest, then the Suzy-slingshot example is not a counterfactual to (C3). For while it is false that had her throwing not happened, the breaking would not have happened (since had she not thrown the rock she would have launched it with her slingshot), it is true that had her throwing not happened and had she been in her default state, the breaking would not have happened.

However, I do not see that Paul and Hall's proposal evades the Suzy-slingshot problem in all its forms. Couldn't Suzy's default state be one in which she is launching the rock with her slingshot? (For terminological convenience I will suppose that there is a state S that one is in iff one is launching a rock with one's slingshot—I believe that Paul and Hall would accept this supposition—even though officially

[32] This isn't exactly how Paul and Hall intend their proposal. Their proposal is not to put the stuff about default states into the antecedent of the conditional, but to interpret the conditional so that if C hadn't happened, the system to which C pertains would have been in its default state. On their interpretation, "If Suzy's throw hadn't happened, the breaking of the window wouldn't have happened" is true even in the slingshot example. I suspect that this is not a meaning the conditional may have in ordinary English; but Paul and Hall are happy to use the conditional in a technical sense for the purpose of producing a theory of causation.

on my theory of states there is not. Indeed, Maudlin's actual distinction is between default (or inertial) and deviant *behavior* (2004, 431), so to the extent that Paul and Hall want their distinction to match Maudlin's, they should recognize the existence of states of behaving in various ways.) Suppose that due to a psychological compulsion Suzy constantly launches rocks with her slingshot. It is only with an immense act of will that she can resist this compulsion. Here she is in front of the window, and she feels the compulsion to launch a rock at the window. But in this case she resists, and instead throws a rock, breaking the window. Does the fact of her compulsion make her default state one in which she launches the rock? Then if her throwing hadn't happened, and she had been in her default state, the window still would have broken. Still, the throw caused the breaking, so this is a counterexample to (C3).

The main heuristic Maudlin gives us for distinguishing default from deviant states is this: a thing's default state is the state it's in when nothing is interfering with it (actually he writes "inertial laws... describe how some entities behave when nothing acts on them" (431)). Well, in this case, launching rocks with a slingshot is the state Suzy is in when nothing is interfering with her.[33]

If you think that her compulsion is interfering with her, since having such a compulsion is an abnormal thing for a human being to have, imagine her to be not human but of some alien species for which it is normal.

If the example is becoming too far-fetched, there are less far-fetched ones, for example involving sharks. Sharks cannot maintain their depth without swimming. If they stop swimming, they sink to the bottom. Some sharks also cannot breathe (extract oxygen from water

[33] One might say that Suzy's default state is the state she's in when her neurons are in their default state. (Or one might pick on the atoms that make her up rather than her neurons.) But Paul and Hall cannot consistently say this. If the default state of a neuron is dormancy, a state in which it is not firing, as Paul and Hall suggest elsewhere, then this proposal entails that Suzy when in her default state is dead. But Paul and Hall say that when in her default state Suzy is very much alive, just not causing any rocks to move.

with their gills) without swimming. Such sharks continue swimming even when they are asleep. For a shark like that, its default state is certainly one in which it is swimming. That's what it does when nothing is interfering with it; swimming is the inertial behavior of sharks. Suppose a shark, while swimming toward a submerged window, punts a rock with its nose at the window, causing it to break. Consider the counterfactual: if the punting of the rock had not happened, and the shark had been in its default state, then the breaking would not have happened. This counterfactual is false. Had the shark been in its default state, it would have been swimming, and since it was swimming toward the window, it would have rammed the window and thereby broken it. Nevertheless, the punting caused the breaking; this is a counterexample to (C3). (If you worry that the shark would have seen the window and avoided it, suppose the shark to be blind; if you worry about how a blind shark punted a rock toward the window, suppose it was luck.)

We can improve on (C2) and (C3) if we stop thinking in event-causal terms and switch to thinking in agent-causal terms. First we need to go back to square one. What does the simple counterfactual theory of causation look like, when stated in agent-causal terms? Like this:

(C1*) X caused Y to Z by Ving iff had X not Ved, Y would not have Zed.

Suzy and her sligshot are a problem for (C1*) as much as for (P2). But in agent-causal terms the problem looks different. Instead of trying to find an event-description to go in for "C*" in "If C had not happened and C* had happened in its place, E would not have happened," we need to find a non-stative verb to go in for "ϕed" in "If X had not Ved but had ϕed instead, Y would not have Zed." Now when we were thinking in event-causal terms, the "null hypothesis," that "C*" is replaced by "nothing," that C is to be excised from history with no replacement, went nowhere. Now, though, we are thinking in agent-causal terms, and the null hypothesis is much more promising. Here the null hypothesis is that instead of Ving X is doing nothing:

(C4) X caused Y to Z by Ving iff, instead of Ving, X had done nothing, Y would not have Zed.

The examples that made trouble for (C2) through (C3) now fall away. Maybe, if Suzy hadn't thrown the rock she would have launched it with her slingshot; still, if she hadn't thrown the rock but instead had done nothing, the window would not have broken. Maybe launching rocks with slingshots is Suzy's "default behavior"; it doesn't matter. The counterfactual in (C4) does not look to what would have happened if Suzy had been engaged in her default behavior; it looks to what would have happened if she had been doing nothing. Similarly for the shark that punts the rock: if it hadn't punted the rock, and instead had been doing nothing, it wouldn't have been swimming, so would not have broken the window by swimming into it.

Paul and Hall's proposal actually steals some of its appeal from (C4). They say that their proposal handles cases like Suzy and the slingshot because Suzy's default state is one in which she is doing nothing (51). It may well be true that for many people their default state is one in which they are doing nothing. But surely it is possible for someone's brain to be so-wired that their default state is one in which they are launching rocks with slingshots. Then the person's default state, and the state in which they are doing nothing, come apart. And (I claim) my theory gets cases like that right, and Paul and Hall's gets them wrong.

IX

If agent causation is basic, we can explain why events but not states can be causes, explain why only certain properties of an event can be properties in virtue of which it is a cause, say what causes what in cases of causation by omission without recognizing the existence of negative events, and formulate counterfactual theories of causation so that Suzy, who would have launched her rock with a slingshot had she not thrown it, is not a counterexample. Together I claim that these constitute good reasons to think that agent causation is basic.

References

Anscombe, G. E. M. 1957. *Intention.* Blackwell.

Bennett, Jonathan 1988. *Events and their Names.* Hackett.

Bishop, Jones 1983. "Agent Causation." *Mind* 92: 61–79.

Brown, Harvey 2005. *Physical Relativity.* Oxford University Press.

Chisholm, Roderick 1976. "The Agent as Cause." In Myles Brand and Douglas Walton (eds.), *Action Theory.* D. Reidel 199–211.

Collins, John, Ned Hall, and L. A. Paul (eds.) 2004. *Causation and Counterfactuals.* MIT Press.

Danto, Arthur 1965. "Basic Actions." *American Philosophical Quarterly* 2: 141–8.

Davidson, Donald 1967. "Causal Relations." *The Journal of Philosophy* 64: 691–703.

Galton, Antony 1984. *The Logic of Aspect.* Oxford University Press.

Hitchcock, Christopher 2001. "The Intransitivity of Causation Revealed in Equations and Graphs." *The Journal of Philosophy* 98: 273–99.

Huddleston, Rodney 2002. "Non-finite and Verbless Clauses." In *The Cambridge Grammar of the English Language.* Cambridge University Press, 1171–272.

Kim, Jaegwon 1999. "Hempel, Explanation, Metaphysics." *Philosophical Studies* 94: 1–20.

Lewis, David 1973. *Counterfactuals.* Blackwell.

Lewis, David 1986a. "Causation." In *Philosophical Papers Volume II.* Oxford University Press, 159–213.

Lewis, David 1986b. "Veridical Hallucination and Prosthetic Vision." In *Philosophical Papers Volume II.* Oxford University Press, 273–90.

Lewis, David 2004. "Causation as Influence." In Collins et al. 2004, 75–106.

Lowe, E. J. 2002. *A Survey of Metaphysics.* Oxford University Press.

Maudlin, Tim 2004. "Causation, Counterfactuals, and the Third Factor." In Collins et al. 2004, 419–43.

McGrath, Sarah 2005. "Causation by Omission: A Dilemma." *Philosophical Studies* 123: 125–58.

Mellor, D. H. 2004. "For Facts as Causes and Effects." In Collins et al. 2004, 309–24.

O'Connor, Timothy 1995. "Agent Causation." In Timothy O'Connor (ed.), *Agents, Causes, and Events: Essays on Indeterminism and Free Will.* Oxford University Press, 173–200.

Parsons, Terence 1990. *Events in the Semantics of English.* MIT Press.

Paul, L. A. and Ned Hall 2013. *Causation: A User's Guide.* Oxford University Press.

Schaffer, Jonathan 2005. "Contrastive Causation." *The Philosophical Review* 114: 327–58.

Setiya, Kieran 2012. "Knowing How." *Proceedings of the Aristotelian Society* 112: 185–307.

Sider, Theodore 2011. *Writing the Book of the World.* Oxford University Press.

Stalnaker, Robert 1968. "A Theory of Conditionals." In W. L. Harper, R. Stalnaker, and G. Pearce (eds.), *Ifs.* Basil Blackwell, 41–55.

Taylor, Richard 1966. *Action and Purpose.* Prentice Hall.

Thompson, Michael 2008. *Life and Action.* Harvard University Press.

Thomson, Judith Jarvis 2003. "Causation: Omissions." *Philosophy and Phenomenological Research* 66: 81–103.

Van Cleve, James 2015. *Problems from Reid.* Oxford University Press.

Yablo, Stephen 2010. "Advertisement for a Sketch of an Outline of a Prototheory of Causation." In *Things.* Oxford University Press, 98–116.

Zwicky, A. M. and J. M. Sadock 1975. "Ambiguity Tests and How to Fail Them." *Syntax and Semantics* 4: 1–36.

Index